Electronics, Controls, and Communications Practice Exam

Second Edition

John A. Camara, PE

PPI®
PPI2PASS.COM
A **KAPLAN** COMPANY

Register Your Book at ppi2pass.com

- Receive the latest exam news.
- Obtain exclusive exam tips and strategies.
- Receive special discounts.

Report Errors for This Book

PPI is grateful to every reader who notifies us of a possible error. Your feedback allows us to improve the quality and accuracy of our products. Report errata at **ppi2pass.com**.

Digital Book Notice

All digital content, regardless of delivery method, is protected by U.S. copyright laws. Access to digital content is limited to the original user/assignee and is non-transferable. PPI may, at its option, revoke access or pursue damages if a user violates copyright law or PPI's end-user license agreement.

ELECTRONICS, CONTROLS, AND COMMUNICATIONS PRACTICE EXAM
Second Edition

Current release of this edition: 1

Release History

date	edition number	revision number	update
Feb 2018	1	5	Minor corrections. Minor cover updates.
Jul 2018	1	6	Minor corrections. Minor formatting and pagination changes.
May 2019	2	1	New edition.

© 2019 Kaplan, Inc. All rights reserved.

All content is copyrighted by Kaplan, Inc. No part, either text or image, may be used for any purpose other than personal use. Reproduction, modification, storage in a retrieval system or retransmission, in any form or by any means, electronic, mechanical, or otherwise, for reasons other than personal use, without prior written permission from the publisher is strictly prohibited. For written permission, contact permissions@ppi2pass.com.

Printed in the United States of America.

PPI
1250 Fifth Avenue, Belmont, CA 94002
(650) 593-9119
ppi2pass.com

ISBN: 978-1-59126-644-0

F E D C B A

Table of Contents

PREFACE AND ACKNOWLEDGMENTS v

CODES AND REFERENCES vii

INTRODUCTION ix

MORNING SESSION 1
    Morning Session Instructions 1
    Morning Session Exam 3

AFTERNOON SESSION 14
    Afternoon Session Instructions 15
    Afternoon Session Exam 17

ANSWER KEYS 27

SOLUTIONS 29
    Morning Session Exam 29
    Afternoon Session Exam 41

Preface and Acknowledgments

The *Electronics, Controls, and Communications Practice Exam* is a practice exam designed to the format and specifications defined by the National Council of Examiners for Engineering and Surveying (NCEES) for the Principles and Practice of Engineering (PE) Electrical and Computer: Electronics, Controls, and Communications exam.

The sample exam provides an opportunity for comprehensive exam preparation: the incorrect answers can enlighten; the correct answers are thoroughly explained; and since the problems are based on a range of electrical and computer engineering topics, you will recognize areas where further study and preparation are needed. With diligent study, you will find the preparation you seek within these pages. At the very least, the questions will lead you to an intelligent search for information. I hope this book serves you well and that you enjoy the adventure of learning.

Many sources were used to define the scope of this book, including NCEES publications pertaining to the NCEES PE Electrical and Computer: Electronics, Controls, and Communications exam, many references listed in the Codes and References section of this book, and survey comments from recent PE examinees.

Should you find an error in this book, know that it is mine, and that I regret it. Beyond that, I hope two things happen. First, please let me know about the error by using the error reporting form on the PPI website, found at **ppi2pass.com**. Second, I hope you learn something from the error—I know I will! I appreciate suggestions for improvement, additional questions, and recommendations for expansion so that new editions or similar texts will better meet the needs of future examinees.

The *Electronics, Controls, and Communications Practice Exam* descended from the third edition of the *Electrical Engineering Sample Examinations*. Thanks go to the very professional and dedicated team at PPI that included Megan Synnestvedt, senior product manager; Steve Shea, content specialist; Beth Christmas, project manager; Scott Marley, senior copy editor; Tom Bergstrom technical drawings and cover designer; Richard Iriye, typesetter; Sam Webster, product data manager; Cathy Schrott, editorial operations; and Grace Wong, director of editorial operations. Thanks also go to Anil Acharya who performed the calculation checks.

A very special thanks to my son, Jac Camara. Our discussions about quantum mechanics and a variety of other topics helped stir the intellectual curiosity I thought was fading.

Thanks also to my daughter, Cassiopeia, who has been a constant source of inspiration, challenge, and pride.

Finally, thanks to my mother, Arlene, who made me a better person.

John A. Camara, PE

Codes and References

The information that was used to write and update this book was based on the exam specifications at the time of publication. However, as with engineering practice itself, the PE exam is not always based on the most current codes or cutting-edge technology. Similarly, codes, standards, and regulations adopted by state and local agencies often lag issuance by several years. It is likely that the codes that are most current, the codes that you use in practice, and the codes that are the basis of your exam will all be different. However, differences between code editions typically minimally affect the technical accuracy of this book, and the methodology presented remains valid. For more information about the variety of codes related to electrical engineering, refer to the following organizations and their websites.

- American National Standards Institute (ansi.org)
- Electronic Components Industry Association (ecianow.org)
- Federal Communications Commission (fcc.gov)
- Institute of Electrical and Electronics Engineers (ieee.org)
- International Organization for Standardization (iso.org)
- International Society of Automation (isa.org)
- National Electrical Manufacturers Association (nema.org)
- National Fire Protection Association (nfpa.org)

The PPI website (**ppi2pass.com**) provides the dates and editions of the codes, standards, and regulations on which NCEES has announced the PE exams are based. It is your responsibility to find out which codes are relevant to your exam.

The minimum recommended library of materials to bring to the PE Electrical and Computer exam consists of this book, any applicable code books, a standard handbook of electrical engineering, and one or two textbooks that cover fundamental circuit theory (both electrical and electronic).

CODES

47 CFR 73: *Code of Federal Regulations*, "Title 47—Telecommunication, Part 73—Radio Broadcast Services," 2018. Office of the Federal Register National Archives and Records Administration, Washington, DC.

IEEE Std 242 (IEEE Buff Book): *IEEE Recommended Practice for Protection and Coordination of Industrial and Commercial Power Systems*, 2001.

IEEE Std 446 (IEEE Orange Book): *IEEE Recommended Practice for Emergency and Standby Power Systems for Industrial and Commercial Applications*, 1995.

IEEE Std 493 (IEEE Gold Book): *IEEE Recommended Practice for the Design of Reliable Industrial and Commercial Power Systems*, 2007.

IEEE Std 602 (IEEE White Book): *IEEE Recommended Practice for Electric Systems in Health Care Facilities*, 2007.

IEEE Std 739 (IEEE Bronze Book): *IEEE Recommended Practice for Energy Management in Industrial and Commercial Facilities*, 1995.

IEEE Std 902 (IEEE Yellow Book): *IEEE Guide for Maintenance, Operation, and Safety of Industrial and Commercial Power Systems*, 1998.

IEEE Std 1015 (IEEE Blue Book): *IEEE Recommended Practice for Applying Low-Voltage Circuit Breakers Used in Industrial and Commercial Power Systems*, 2006.

NEC (NFPA 70): *National Electrical Code*, 2017. National Fire Protection Association, Quincy, MA.

NESC (IEEE C2): 2017 *National Electrical Safety Code*, 2017. The Institute of Electrical and Electronics Engineers, Inc., New York, NY.

REFERENCES

An Introduction to Digital and Analog Integrated Circuits and Applications. Sanjit K. Mitra. Harper & Row, Publishers. [Digital Circuit Fundamentals textbook]

Applied Electromagnetics. Martin A. Plonus. McGraw-Hill. [Electromagnetic Theory textbook]

CRC Materials Science and Engineering Handbook. James F. Shackelford and William Alexander, eds. CRC Press, Inc. [General engineering handbook]

CRC Standard Mathematical Tables and Formulae. William H. Beyer, ed. CRC Press, Inc. [General engineering reference]

Electronics Engineers' Handbook. Donald Christiansen, ed. McGraw-Hill. [Electrical and Electronics handbook]

Introduction to Computer Engineering: Hardware and Software Design. Taylor L. Booth. John Wiley & Sons. [Computer Design Basics textbook]

Linear Circuits. M.E. Van Valkenberg and B.K. Kinariwala. Prentice Hall, Inc. [AC/DC Fundamentals textbook]

McGraw-Hill Dictionary of Scientific and Technical Terms. Sybil P. Parker, ed. McGraw-Hill. [General engineering reference]

McGraw-Hill Internetworking Handbook. Ed Taylor. McGraw-Hill. [Computer handbook]

Microelectronic Circuit Design. Richard C. Jaeger and Travis Blalock. McGraw-Hill [Electronic Fundamentals textbook]

Microelectronics. Jacob Millman and Arvin Millman. McGraw-Hill. [Electronic Fundamentals textbook]

National Electrical Code Handbook. National Fire Protection Association. [Power handbook]

Process/Industrial Instruments and Controls Handbook. Gregory K. McMillan and Douglas Considine, eds. McGraw-Hill. [Power and Electrical and Electronics handbook]

Standard Handbook for Electrical Engineers. Donald G. Fink and H. Wayne Beaty. McGraw-Hill. [Power and Electrical and Electronics handbook]

The Biomedical Engineering Handbook. Joseph D. Brozino, ed. CRC Press. [Electrical and Electronics handbook]

The Communications Handbook. Jerry D. Gibson, ed. CRC Press, Inc. [Electrical and Electronics handbook]

The Computer Science and Engineering Handbook. Allen B. Tucker, Jr., ed. CRC Press, Inc. [Computer reference]

The Internet for Scientists and Engineers, Brian J. Thomas. SPIE Press. [Computer reference]

Wireless and Cellular Communications. William C.Y. Lee. McGraw-Hill. [Electrical and Electronics handbook]

Introduction

ABOUT THE NCEES PE ELECTRICAL AND COMPUTER: ELECTRONICS, CONTROLS, AND COMMUNICATIONS EXAM

The NCEES PE Electrical and Computer: Electronics, Controls, and Communications exam is made up of 80 problems and is divided into two four-hour long sessions. The format of all exam problems is multiple-choice with a problem statement and all required defining information, followed by four logical choices. Only one of the four options is correct, and the problems are completely independent of each other.

The topics and the distribution of problems for the NCEES PE Electronics, Controls, and Communications exam are as follows. The distribution both on this practice exam and on the NCEES exam is approximate.

NCEES PE Electrical and Computer: Electronics, Controls, and Communications Exam

- General Electrical Engineering Knowledge (32 questions):

  circuit analysis; measurement and instrumentation; safety and reliability; signal processing

- Digital Systems (8 questions):

  digital logic; digital components

- Electromagnetics (8 questions):

  electromagnetic fields; guided waves; antennas

- Electronics (16 questions):

  electronics circuits; electronic components and applications

- Control Systems (8 questions):

  analysis and design of analog or digital control systems

- Communications (8 questions):

  modulation techniques; noise and interference; communication systems

According to the NCEES, exam questions related to codes and standards will be based on either (1) an interpretation of a code or standard that is presented in the exam booklet or (2) a code or standard that a committee of licensed engineers feels minimally competent engineers should know.

For further information about the exams, and tips on how to prepare for the exams, consult PPI's website, **ppi2pass.com**.

HOW TO USE THIS BOOK

Prior to taking the practice exam in this book, assemble your materials as if you are taking the actual exam. Obtain a copy of any reference or code books you plan to use on the exam. (Use the Codes and References section to determine which supplementary materials you will need for the exam.) Since some states have restrictions on the materials you are allowed to use during the exam, visit **ppi2pass.com** to find a link to your state's board of engineering and check for any restrictions. Follow these restrictions when taking the practice exam. Remember that adequate preparation, not an extensive library, is the key to success, both when you take the practice exam and the actual exam.

After you feel that you have sufficiently prepared for the practice exam in this book, read the instructions at the beginning of the exam for guidance on how to properly simulate the exam. Set a timer for four hours, and take the morning session. After a one-hour break, turn to the afternoon exam, set the timer, and complete the afternoon session.

Next, check your answers and read through the solutions of the problems that you answered incorrectly or were unable to answer. Evaluate your strengths and weaknesses and select additional texts to supplement your studies. Check the PPI website at **ppi2pass.com** for the latest in exam preparation materials.

The keys to exam success are knowing the basics and practicing solving as many problems as possible. This book will assist you with both objectives.

Morning Session Instructions

In accordance with the rules established by your state, you may use textbooks, handbooks, bound reference materials, and any approved battery- or solar-powered, silent calculator to work this examination. However, no blank papers, writing tablets, unbound scratch paper, or loose notes are permitted. Sufficient room for scratch work is provided in the Examination Booklet.

You are not permitted to share or exchange materials with other examinees. However, the books and other resources used in this morning session may be changed prior to the afternoon session.

You will have four hours in which to work this session of the examination. Your score will be determined by the number of questions that you answer correctly. There is a total of 40 questions. All 40 questions must be worked correctly in order to receive full credit on the exam. There are no optional questions. Each question is worth 1 point. The maximum possible score for this section of the examination is 40 points.

Partial credit is not available. No credit will be given for methodology, assumptions, or work written in your Examination Booklet.

Record all of your answers on the Answer Sheet. No credit will be given for answers marked in the Examination Booklet. Mark your answers with a no. 2 pencil. Answers marked in pen may not be graded correctly. Marks must be dark and must completely fill the bubbles. Record only one answer per question. If you mark more than one answer, you will not receive credit for the question. If you change an answer, be sure the old bubble is erased completely; incomplete erasures may be misinterpreted as answers.

If you finish early, check your work and make sure that you have followed all instructions. After checking your answers, you may turn in your Examination Booklet and Answer Sheet and leave the examination room. Once you leave, you will not be permitted to return to work or change your answers.

When permission has been given by your proctor, break the seal on the Examination Booklet. Check that all pages are present and legible. If any part of your Examination Booklet is missing, your proctor will issue you a new Booklet.

Do not work any questions from the Afternoon Session during the first four hours of this exam.

WAIT FOR PERMISSION TO BEGIN

Name: _____
     Last   First   Middle Initial

Examinee number: _____

Examination Booklet number: _____

Principles and Practice of Engineering Examination

Morning Session Practice Examination

ELECTRONICS, CONTROLS, AND COMMUNICATIONS PRACTICE EXAM

1. Ⓐ Ⓑ Ⓒ Ⓓ
2. Ⓐ Ⓑ Ⓒ Ⓓ
3. Ⓐ Ⓑ Ⓒ Ⓓ
4. Ⓐ Ⓑ Ⓒ Ⓓ
5. Ⓐ Ⓑ Ⓒ Ⓓ
6. Ⓐ Ⓑ Ⓒ Ⓓ
7. Ⓐ Ⓑ Ⓒ Ⓓ
8. Ⓐ Ⓑ Ⓒ Ⓓ
9. Ⓐ Ⓑ Ⓒ Ⓓ
10. Ⓐ Ⓑ Ⓒ Ⓓ
11. Ⓐ Ⓑ Ⓒ Ⓓ
12. Ⓐ Ⓑ Ⓒ Ⓓ
13. Ⓐ Ⓑ Ⓒ Ⓓ
14. Ⓐ Ⓑ Ⓒ Ⓓ
15. Ⓐ Ⓑ Ⓒ Ⓓ
16. Ⓐ Ⓑ Ⓒ Ⓓ
17. Ⓐ Ⓑ Ⓒ Ⓓ
18. Ⓐ Ⓑ Ⓒ Ⓓ
19. Ⓐ Ⓑ Ⓒ Ⓓ
20. Ⓐ Ⓑ Ⓒ Ⓓ
21. Ⓐ Ⓑ Ⓒ Ⓓ
22. Ⓐ Ⓑ Ⓒ Ⓓ
23. Ⓐ Ⓑ Ⓒ Ⓓ
24. Ⓐ Ⓑ Ⓒ Ⓓ
25. Ⓐ Ⓑ Ⓒ Ⓓ
26. Ⓐ Ⓑ Ⓒ Ⓓ
27. Ⓐ Ⓑ Ⓒ Ⓓ
28. Ⓐ Ⓑ Ⓒ Ⓓ
29. Ⓐ Ⓑ Ⓒ Ⓓ
30. Ⓐ Ⓑ Ⓒ Ⓓ
31. Ⓐ Ⓑ Ⓒ Ⓓ
32. Ⓐ Ⓑ Ⓒ Ⓓ
33. Ⓐ Ⓑ Ⓒ Ⓓ
34. Ⓐ Ⓑ Ⓒ Ⓓ
35. Ⓐ Ⓑ Ⓒ Ⓓ
36. Ⓐ Ⓑ Ⓒ Ⓓ
37. Ⓐ Ⓑ Ⓒ Ⓓ
38. Ⓐ Ⓑ Ⓒ Ⓓ
39. Ⓐ Ⓑ Ⓒ Ⓓ
40. Ⓐ Ⓑ Ⓒ Ⓓ

Morning Session Exam

1. A schematic and test data sheet for a low-dropout linear voltage regulator are shown (see *Test Data Sheet for Problem 1*). The input voltage to be regulated is 5 V. The minimum output current required to maintain voltage regulation is 1.7 A.

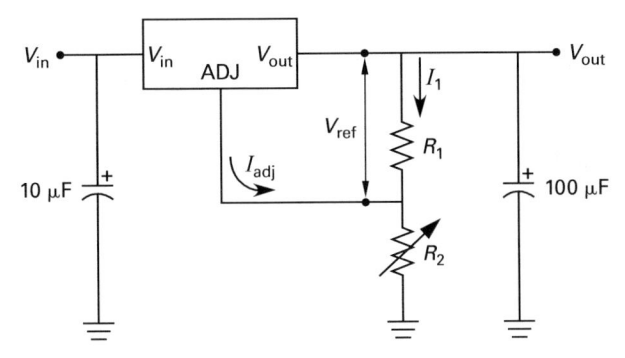

$$V_{out} = V_{ref}\left(1 + \frac{R_2}{R_1}\right) + I_{adj}R_2$$

A load attached to the output is not working as expected. During troubleshooting, the current from the V_{out} terminal is measured as 800 mA. The expected output voltage, V_{out}, is most nearly

- (A) 1.3 V
- (B) 1.8 V
- (C) 3.3 V
- (D) 5.0 V

2. Five paths are indicated in the circuit shown.

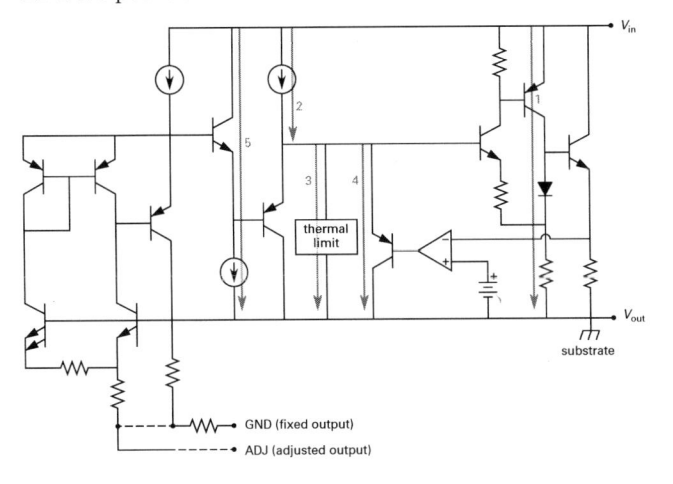

Which two paths result in current limiting?

- (A) 1 and 5
- (B) 2 and 3
- (C) 2 and 4
- (D) 2 and 5

3. A control system uses software-based feedback in a computer as shown.

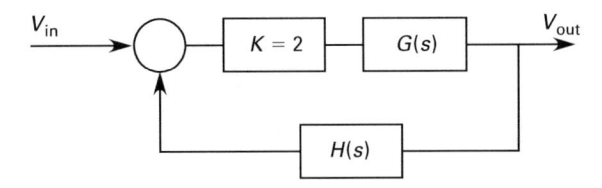

The gain is 2, and the process is given by the first-order model.

$$G(s) = \frac{80}{1 + (30 \text{ s})s}$$

The feedback provided by the computer is accurate but delayed by t_d.

$$H(s) = e^{-t_d s}$$

The maximum delay time (processing time) that can be allowed to determine the closed-loop stability is most nearly

- (A) 0.30 s
- (B) 5.3 s
- (C) 28 s
- (D) 89 s

Test Data Sheet for Problem 1

parameter		test conditions	min	typ	max	unit
V_{ref} reference voltage		$I_{out} = 10$ mA, $V_{in} - V_{out} = 2$ V, $T_J = 25°C$	1.238	1.25	1.262	V
	10 mA $\leq I_{out} \leq$ 800 mA, 1.4 V $\leq V_{in} - V_{out} \leq$ 10 V	$T_J = 25°C$		1.25		
		over the junction temperature 0°C to 125°C	1.225		1.27	
V_{out} output voltage		$I_{out} = 10$ mA, $V_{in} = 3.8$ V, $T_J = 25°C$	1.782	1.8	1.818	V
	0 mA $\leq I_{out} \leq$ 800 mA, 3.2 V $\leq V_{in} \leq$ 10 V	$T_J = 25°C$		1.8		
		over the junction temperature 0°C to 125°C	1.746		1.854	
		$I_{out} = 10$ mA, $V_{in} = 4.5$ V, $T_J = 25°C$	2.475	2.5	2.525	V
	0 mA $\leq I_{out} \leq$ 800 mA, 3.9 V $\leq V_{in} \leq$ 10 V	$T_J = 25°C$		2.5		
		over the junction temperature 0°C to 125°C	2.45		2.55	
		$I_{out} = 10$ mA, $V_{in} = 5$ V, $T_J = 25°C$	3.267	3.3	3.333	V
	0 mA $\leq I_{out} \leq$ 800 mA, 4.75 V $\leq V_{in} \leq$ 10 V	$T_J = 25°C$		3.3		
		over the junction temperature 0°C to 125°C	3.235		3.365	
		$I_{out} = 10$ mA, $V_{in} = 7$ V, $T_J = 25°C$	4.95	5.0	5.05	V
	0 mA $\leq I_{out} \leq$ 800 mA, 6.5 V $\leq V_{in} \leq$ 10 V	$T_J = 25°C$		5.0		
		over the junction temperature 0°C to 125°C	4.9		5.1	

Courtesy of Texas Instruments

4. What time-domain function is associated with the given Laplace transform?

$$F(s) = \frac{2}{s^3(s-3)}$$

(A) $t^2 e^3$

(B) $t^2(1 - e^{3t})$

(C) $-\frac{2}{9}e^{3t} + t^2 + \frac{t}{3} + \frac{2}{9}$

(D) $\frac{2}{27}e^{3t} - \frac{1}{3}t^2 - \frac{2}{9}t - \frac{2}{27}$

5. For the circuit shown, the input voltage, v_{in}, is $(1 \text{ V})\sin 377t$. The output voltage is 0 V at $t = 0$ s.

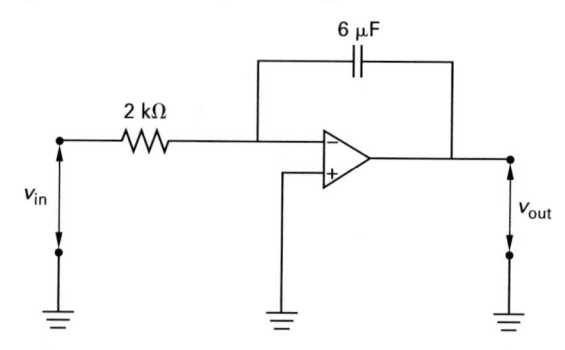

Assuming ideal conditions, what is the expression for the output voltage of the circuit when $t \geq 0$ s?

(A) $(0.221 \text{ V})\cos(377t - 1)$

(B) $(0.221 \text{ V})\cos 377t$

(C) $(1.13 \text{ V})\sin 377t$

(D) $(1.13 \text{ V})\sin(377t + 1)$

6. For the circuit shown, what is the voltage at node A?

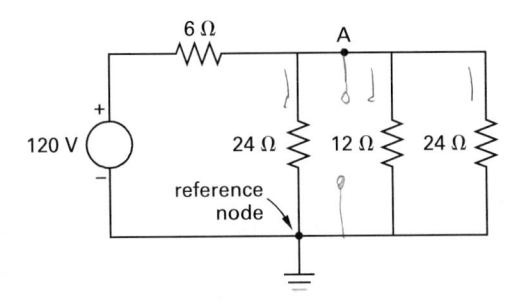

(A) 12 V
(B) 30 V
(C) 60 V
(D) 96 V

7. Which of the following amplifier circuits, designed using a single op amp, would exhibit a 0.75 gain, an 800 kΩ input resistance, and a 0 Ω output resistance?

(A)

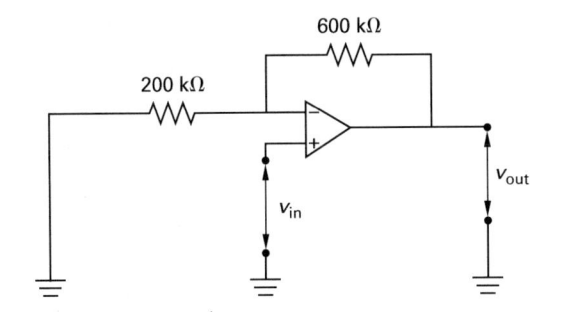

(B)

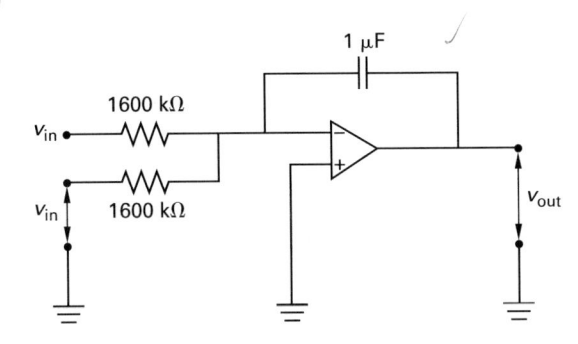

(C)

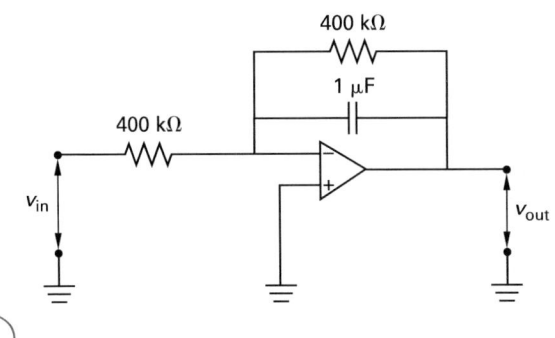

(D)

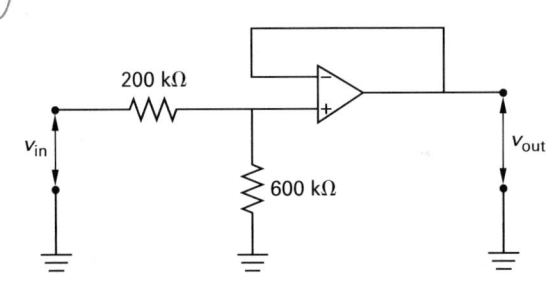

8. The given tuning circuit is used in a communication circuit and the internal resistance of the inductor is 3 Ω.

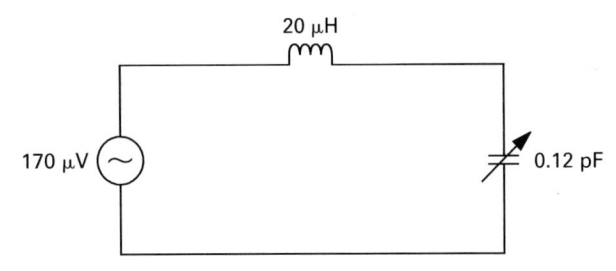

What is most nearly the resonant frequency of the circuit?

(A) 0.15 MHz
(B) 16 MHz
(C) 100 MHz
(D) 250 MHz

9. A communications line uses a shield twisted pair (STP) cable that has a velocity factor of 0.6. What is most nearly the wavelength of a 1 GHz signal on the STP cable?

(A) 0.2 m
(B) 0.3 m
(C) 0.6 m
(D) 1 m

10. A communication circuit senses the following voltage waveform.

$$v(t) = 0.5 \text{ V} + (1.7 \text{ V})\sin(2513t + \theta)$$

The phase angle, θ, is 2 rad. t is measured in seconds. What is most nearly the period of the waveform?

(A) 1.2 ms
(B) 2.5 ms
(C) 25 ms
(D) 400 ms

11. A given function is represented as

$$3e^{-2t}\cos(377t - 30°)$$

Assume that the cosine term will be used as the reference. The periodic portion of the function will be represented by e^{st}. What is the value of **s**?

(A) $-2 + j3$
(B) $\dfrac{\pi}{6} + j3$
(C) $3 + j\left(\dfrac{\pi}{6}\right)$
(D) $-2 + j377$

12. For the circuitry shown, assume near ideal operational amplifier characteristics.

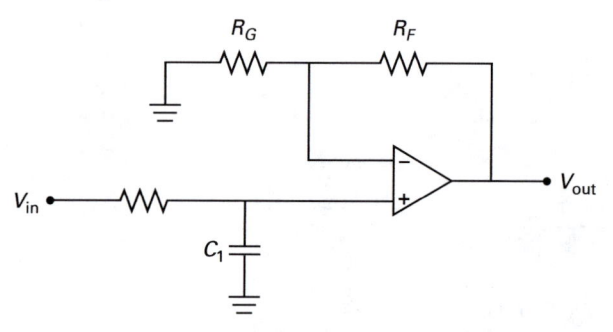

What type of filter is shown?

(A) band pass
(B) low-pass
(C) mid-band
(D) narrow band

13. In the parallel analog-to-digital converter shown, the op amp supply voltages are ± 5 V, and the reference voltage, V_{ref}, is 1 V.

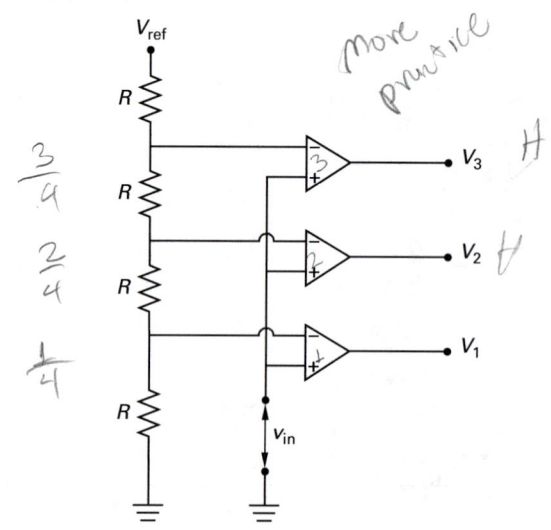

The output voltage is considered high at $+5$ V and low at -5 V. If $v_{\text{in}} = 0.76$ V, what is the output, assuming an output order of V_3, V_2, V_1?

(A) L, L, L
(B) L, L, H
(C) L, H, H
(D) H, H, H

14. For the following op amp configuration, used to provide gain in an electronic circuit, what is the expression for v_{out}?

(A) v_{in}

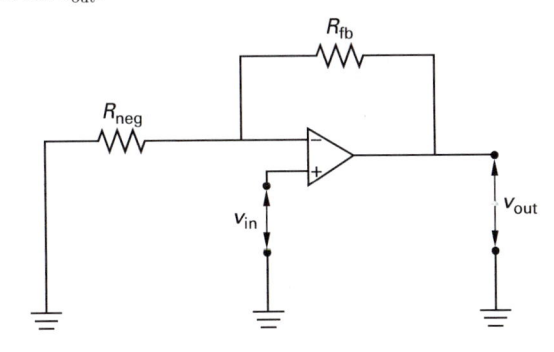

(B) $v_{in}\left(-\dfrac{R_{fb}}{R_{neg}}\right)$

(C) $v_{in}\left(1+\dfrac{R_{fb}}{R_{neg}}\right)$

(D) $\left(\dfrac{R_{fb}}{R_{neg}}\right)(v^+ - v^-)$

15. For the circuit shown, which is being used as a buffer, what is the expression for the output voltage?

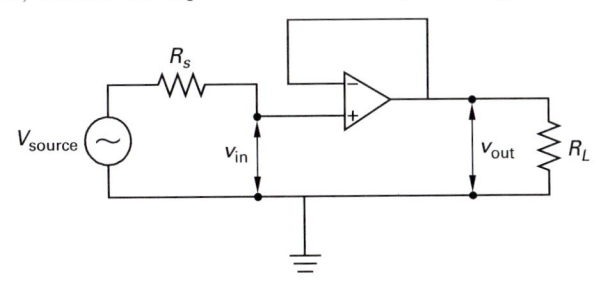

(A) v_{in}

(B) $-R_s v_{in}$

(C) $v_{in}(1+R_s)$

(D) $R_s(v^+ - v^-)$

16. An amplitude modulation (AM) carrier signal equal to $4\sin 100t$ is modulated by the signal $1+3.6\sin 25t$. The modulated spectrum is carried at $+\omega_c$. The power delivered to a 1 Ω resistor at 75 rad/s is most nearly

(A) 5.1 W

(B) 7.2 W

(C) 26 W

(D) 52 W

17. What is the magnitude of the current through the 18 Ω resistor shown?

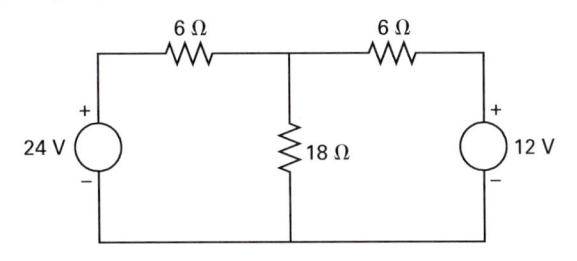

(A) 0 A

(B) 4/25 A

(C) 1/2 A

(D) 6/7 A

more practice

18. The noise figure for an electronic circuit is listed in the manufacturer's data sheet as 1.5 dB. The system in which it will be used is expected to have a signal-to-noise ratio (SNR) of 2.0 dB at the input. The output SNR for the system design is most nearly

(A) −3.5 dB

(B) −1.5 dB

(C) −0.5 dB

(D) 0.5 dB

19. The noise temperature for an integrated circuit (IC) chip is listed as 300K. The noise factor for this chip is most nearly

(A) −0.03

(B) 0.03

(C) 1

(D) 2

20. A capacitance meter reads 15 μF when placed across the terminals of an electrolytic capacitor. A voltmeter across the same terminals reads 90 V. What is most nearly the charge on the capacitor?

(A) 0.14 μC

(B) 14 μC

(C) 1.4 mC

(D) 6.0 mC

21. What is the magnitude of the Norton equivalent current for the circuit shown, with $R = 2\,\Omega$?

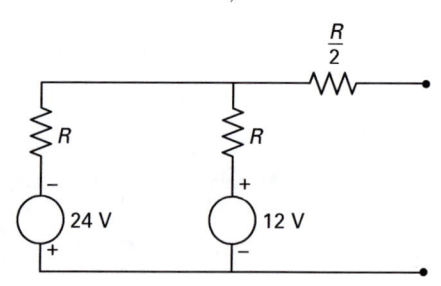

(A) 3.0 A
(B) 6.0 A
(C) 9.0 A
(D) 12 A

22. A simple communication system has a call arrival rate of 90,000 per hour with an average call time of 3 min. The traffic capacity is

(A) 20 erlangs
(B) 1500 erlangs
(C) 4500 erlangs
(D) 270,000 erlangs

23. A control system uses the input filter circuitry shown.

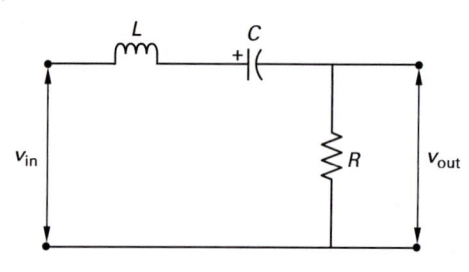

The input impedance normalized to R (i.e., $|Z_{in}/R|$) for various input filter circuitry is shown. Which is the actual frequency response of the input filter circuitry?

(A) 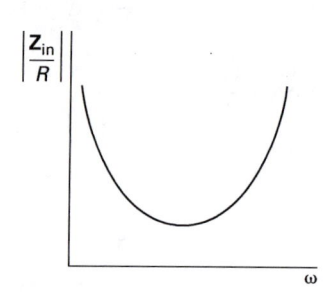
(B) 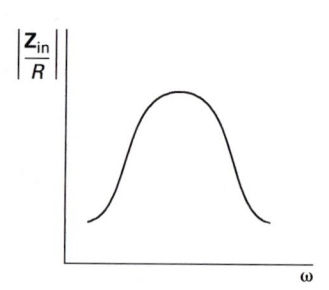
(C) 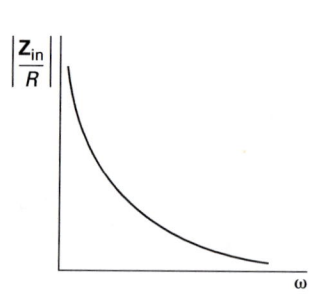
(D)

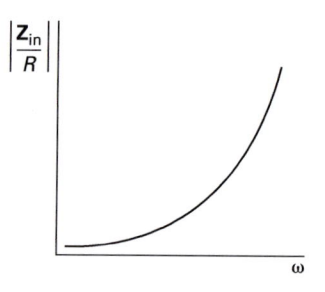

24. The RC circuit shown contains a capacitor with an initial voltage of $v_C(0)$.

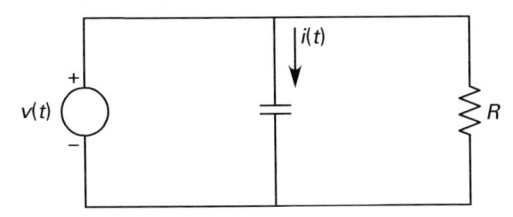

What is the expression for the current $I(s)$?

(A) $sCV(s) - sCv_C(s)$

(B) $C(sV(s) - v_C(0))$

(C) $\dfrac{R}{\dfrac{1}{sC} + R}$

(D) $\dfrac{V(s)}{sC - v_C(s)}$

25. An optical isolator is used to separate a sensor from a control system. Silicon is used with a band gap of 1.1 eV. The control system must be able to respond to the frequency of the sensed light to function properly. The frequency of light used in the optical isolator is most nearly

(A) 170 kHz

(B) 240 MHz

(C) 270 GHz

(D) 270 THz

26. A 555 timer integrated circuit is to be used as part of a microprocessor embedded control system. The design chosen consists of one of the setups and one of the outputs shown.

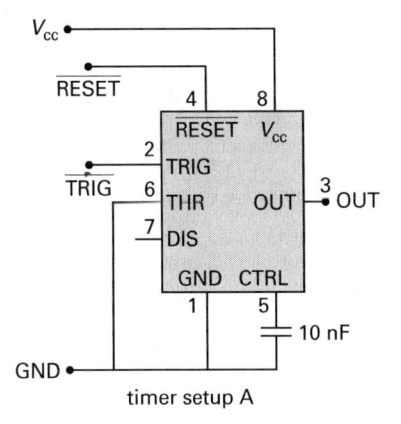

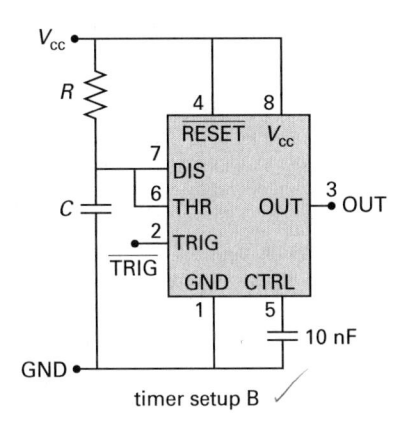

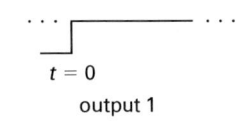

timer setup A

timer setup B

output 1

output 2

The design combination consists of

(A) setup A and output 1

(B) setup A and output 2

(C) setup B and output 1

(D) setup B and output 2

27. A phase-locked loop functional block diagram is shown.

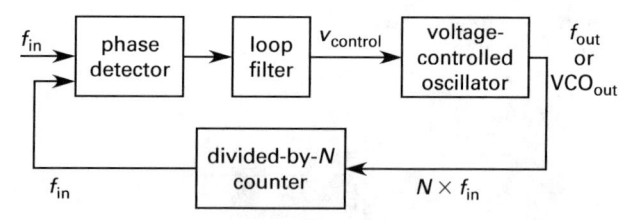

The open loop gain is given by

$$\frac{d\omega_{out}}{d\Delta\phi} = K_{PD}K_{LF}K_{VCO} = K_{PLL}$$

For a given design, the phase detector in its region of operation has a slope of 0.15 V/rad. The loop filter has a gain of 4. The slope of the VCO tuning curve is 2.2 MHz/V. If the input frequency changes by 0.2 rad, the frequency change at the output of the VCO is most nearly

(A) 0.20 MHz
(B) 0.26 MHz
(C) 1.3 MHz
(D) 2.2 MHz

28. The **E** field within a material is given by

$$\mathbf{E} = \left(175 \; \frac{V}{m}\right)\sin 10^9 t$$

The relative permittivity, ϵ_r, is 1. The direction of the field is constant. What is the expression for the magnitude of the displacement current density?

(A) $\left(1.55 \; \frac{A}{m^2}\right)\cos 10^9 t$

(B) $\left(1.55 \; \frac{A}{m^2}\right)\sin 10^9 t$

(C) $\left(1.75 \times 10^2 \; \frac{A}{m^2}\right)\cos 10^9 t$

(D) $\left(1.75 \times 10^{11} \; \frac{A}{m^2}\right)\cos 10^9 t$

29. Which of the following is NOT normally true of a UART?

(A) Data transmission is asynchronous.
(B) No controlling clock is required.
(C) Data packets consist of 5 bits to 9 bits.
(D) The transmission line is held low until the start bit is received.

30. An amplifier consists of multiple stages with gains as shown. A voltage-divider network provides the feedback.

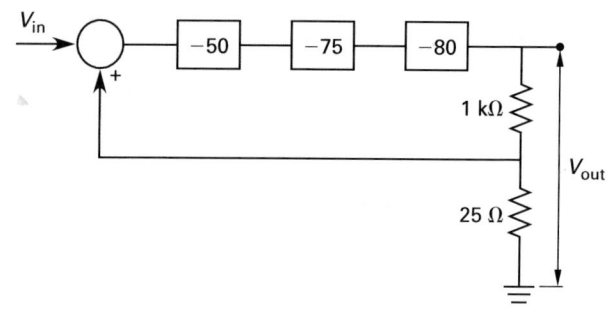

What is most nearly the overall gain of the network, with feedback?

(A) $-300\,000$
(B) -40
(C) 0.03
(D) 40

31. The transfer function of a given filtering network is

$$T(s) = \frac{-10^3}{s^2 + 20s + 10^6}$$

What is the quality factor of the filter?

(A) 20
(B) 50
(C) 50 000
(D) 1 000 000

32. A universal asynchronous receiver/transmitter (UART) uses the midpoint of the data bit as its sampling point to determine the value of the information sent. The period is defined from sampling point to sampling point. The UART is used at a common 9600 baud rate, assuming one bit carries information. The bit period for the UART is most nearly

(A) 100 μs

(B) 960 μs

(C) 9600 μs

(D) 10,000 μs

33. A system is acted upon by a constant force of two units commencing at $t = 0$ s. The transfer function is shown. What is the time-based response function, $r(t)$?

$$T(s) = \frac{6}{s+2}$$

(A) $12e^{-12t}$

(B) $6e^{2t}$

(C) $1 - e^{-2t}$

(D) $6 - 6e^{-2t}$

34. A response function is given by

$$R(s) = \frac{1}{s(s+1)}$$

What is the value of the response function $r(t)$ as t approaches ∞; that is, what is the final steady-state value?

(A) 0

(B) 1

(C) ∞

(D) indeterminate

35. Which of the following statements referring to a controller effect known as quantization is FALSE?

(A) Quantization forces a minimum nonzero change in a measurement for any sampled calculation.

(B) Quantization could also be called resolution.

(C) The control system challenge is an effective derivative action while keeping quantization within the bounds of the accuracy of the data.

(D) The use of a "bucket brigade" algorithm approximates the effects of quantization.

36. The transfer function for a given network is

$$T(s) = \frac{3(s+4)}{(s+2)(s^2+2s+2)}$$

Which of the following illustrations represents the pole-zero diagram for the given transfer function?

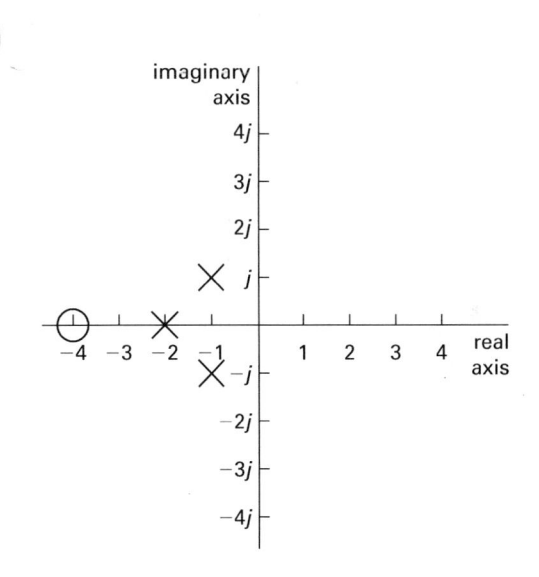

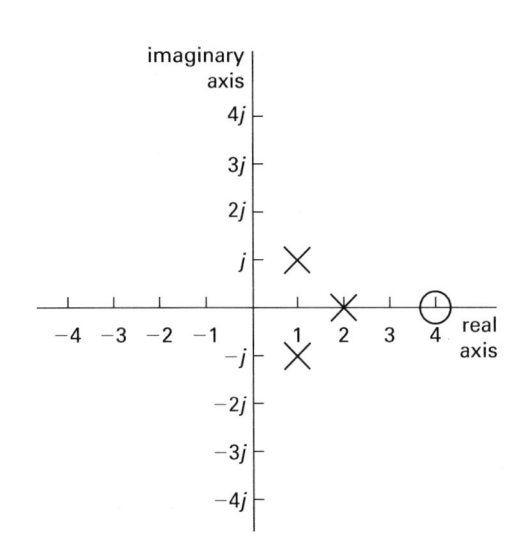

(A)

(B)

(C)

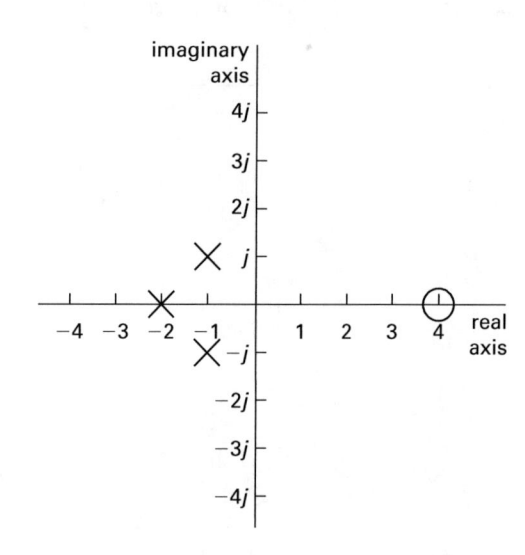

(D)

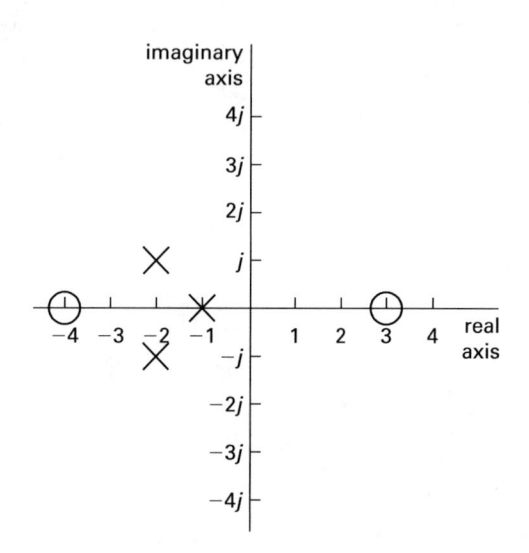

37. Consider the following generic Bode plots of gain and phase angle.

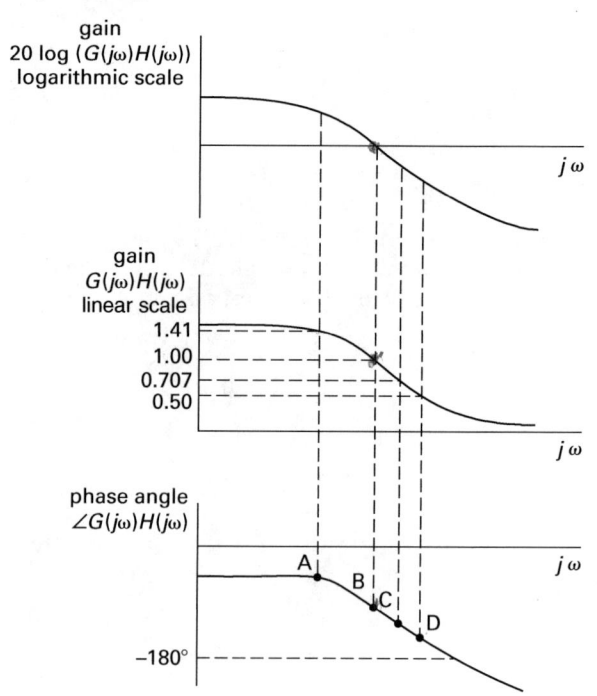

What point represents the phase margin?

(A) point A
(B) point B
(C) point C
(D) point D

38. For a transmission line with a characteristic impedance of 60 Ω and a terminating resistance of 120 Ω, what is the reflection coefficient?

(A) 0.33
(B) 0.50
(C) 2.0
(D) 3.0

39. A space-based antenna transmits at 2 W with a gain of 17 dBW. The path loss is 200 dBW. The receiving antenna has a gain of 50. What is most nearly the effective isotropically radiated power (EIRP) in dBW?

(A) −170 dBW

(B) −110 dBW

(C) 15 dBW

(D) 20 dBW

40. For a modulated FM signal with the following form, what is the approximate maximum frequency deviation from the carrier frequency?

$$s_{\text{mod}}(t) = A\cos(5.5 \times 10^{11} t + 8\sin 10^6 t)$$

(A) 8.8×10^5 Hz

(B) 1.0×10^6 Hz

(C) 1.3×10^6 Hz

(D) 8.0×10^6 Hz

STOP!

DO NOT CONTINUE!

This concludes the Morning Session of the examination. If you finish early, check your work and make sure that you have followed all instructions. After checking your answers, you may turn in your examination booklet and answer sheet and leave the examination room. Once you leave, you will not be permitted to return to work or change your answers.

Afternoon Session Instructions

In accordance with the rules established by your state, you may use textbooks, handbooks, bound reference materials, and any approved battery- or solar-powered, silent calculator to work this examination. However, no blank papers, writing tablets, unbound scratch paper, or loose notes are permitted. Sufficient room for scratch work is provided in the Examination Booklet.

You are not permitted to share or exchange materials with other examinees. However, the books and other resources used in this afternoon session do not have to be the same as were used in the morning session.

You will have four hours in which to work this session of the examination. Your score will be determined by the number of questions that you answer correctly. There is a total of 40 questions. All 40 questions must be worked correctly in order to receive full credit on the exam. There are no optional questions. Each question is worth 1 point. The maximum possible score for this section of the examination is 40 points.

Partial credit is not available. No credit will be given for methodology, assumptions, or work written in your Examination Booklet.

Record all of your answers on the Answer Sheet. No credit will be given for answers marked in the Examination Booklet. Mark your answers with a no. 2 pencil. Answers marked in pen may not be graded correctly. Marks must be dark and must completely fill the bubbles. Record only one answer per question. If you mark more than one answer, you will not receive credit for the question. If you change an answer, be sure the old bubble is erased completely; incomplete erasures may be misinterpreted as answers.

If you finish early, check your work and make sure that you have followed all instructions. After checking your answers, you may turn in your Examination Booklet and Answer Sheet and leave the examination room. Once you leave, you will not be permitted to return to work or change your answers.

When permission has been given by your proctor, break the seal on the Examination Booklet. Check that all pages are present and legible. If any part of your Examination Booklet is missing, your proctor will issue you a new Booklet.

Do not work any questions from the Morning Session during the second four hours of this exam.

WAIT FOR PERMISSION TO BEGIN

Name: _____
       Last   First   Middle Initial

Examinee number: _____

Examination Booklet number: _____

Principles and Practice of Engineering Examination

Afternoon Session Practice Examination

41. Ⓐ Ⓑ Ⓒ Ⓓ	51. Ⓐ Ⓑ Ⓒ Ⓓ	61. Ⓐ Ⓑ Ⓒ Ⓓ	71. Ⓐ Ⓑ Ⓒ Ⓓ
42. Ⓐ Ⓑ Ⓒ Ⓓ	52. Ⓐ Ⓑ Ⓒ Ⓓ	62. Ⓐ Ⓑ Ⓒ Ⓓ	72. Ⓐ Ⓑ Ⓒ Ⓓ
43. Ⓐ Ⓑ Ⓒ Ⓓ	53. Ⓐ Ⓑ Ⓒ Ⓓ	63. Ⓐ Ⓑ Ⓒ Ⓓ	73. Ⓐ Ⓑ Ⓒ Ⓓ
44. Ⓐ Ⓑ Ⓒ Ⓓ	54. Ⓐ Ⓑ Ⓒ Ⓓ	64. Ⓐ Ⓑ Ⓒ Ⓓ	74. Ⓐ Ⓑ Ⓒ Ⓓ
45. Ⓐ Ⓑ Ⓒ Ⓓ	55. Ⓐ Ⓑ Ⓒ Ⓓ	65. Ⓐ Ⓑ Ⓒ Ⓓ	75. Ⓐ Ⓑ Ⓒ Ⓓ
46. Ⓐ Ⓑ Ⓒ Ⓓ	56. Ⓐ Ⓑ Ⓒ Ⓓ	66. Ⓐ Ⓑ Ⓒ Ⓓ	76. Ⓐ Ⓑ Ⓒ Ⓓ
47. Ⓐ Ⓑ Ⓒ Ⓓ	57. Ⓐ Ⓑ Ⓒ Ⓓ	67. Ⓐ Ⓑ Ⓒ Ⓓ	77. Ⓐ Ⓑ Ⓒ Ⓓ
48. Ⓐ Ⓑ Ⓒ Ⓓ	58. Ⓐ Ⓑ Ⓒ Ⓓ	68. Ⓐ Ⓑ Ⓒ Ⓓ	78. Ⓐ Ⓑ Ⓒ Ⓓ
49. Ⓐ Ⓑ Ⓒ Ⓓ	59. Ⓐ Ⓑ Ⓒ Ⓓ	69. Ⓐ Ⓑ Ⓒ Ⓓ	79. Ⓐ Ⓑ Ⓒ Ⓓ
50. Ⓐ Ⓑ Ⓒ Ⓓ	60. Ⓐ Ⓑ Ⓒ Ⓓ	70. Ⓐ Ⓑ Ⓒ Ⓓ	80. Ⓐ Ⓑ Ⓒ Ⓓ

Afternoon Session Exam

41. Consider the electronic circuit models shown. Identifying subscripts have been removed. Which of these models represents a field-effect transistor?

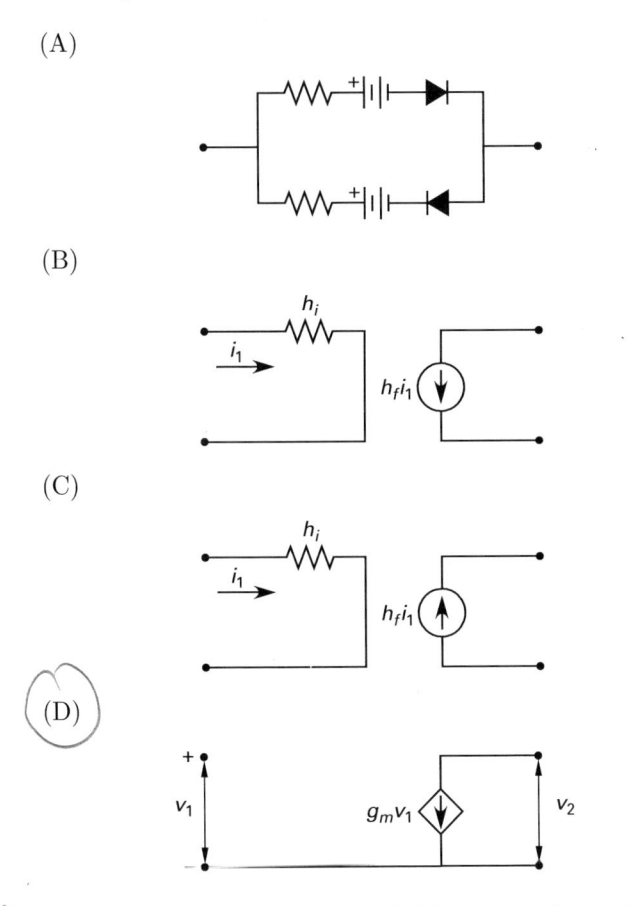

(A)

(B)

(C)

(D)

42. A vibration estimate, provided by a transducer, is transmitted to a system through a series of cascaded amplifiers. The first two amplifier stages are shown. The system, which is controlled by a microprocessor, keeps harmonics below specified values by adjusting the overall process rate through these stages.

For practical purposes, the measurement input is periodic at V_2 and is characterized by the Fourier series

$$V_2(t) = 3.0\sin 377t + 1.5\sin 377t + 0.75\sin 377t + \cdots$$

The system controls the level of the harmonics at V_1. The operational amplifiers may be treated as ideal amplifiers.

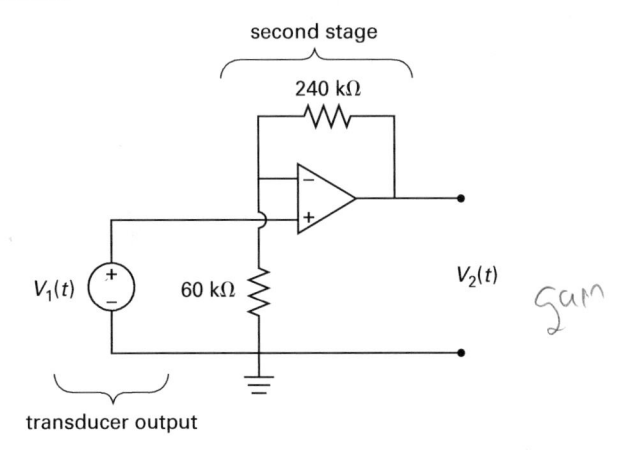

The level of the second harmonic in $V_1(t)$ is

(A) 0.3
(B) 0.5
(C) 1.5
(D) 25

43. The troubleshooting of a measurement system has started at the sensor, which is a linear variable differential transformer (LVDT) used to measure the position of a bulkhead. The LVDT data sheet indicates an output of ± 12 V gives a full-scale deflection. The data sheet further indicates a typical error in a given measurement is 0.5% full scale. For testing, the LVDT is manually placed at the position of 75 engineering units, and the output is measured. The expected range of the output is most nearly

(A) 8.2 V to 9.7 V
(B) 8.4 V to 9.6 V
(C) 8.6 V to 9.4 V
(D) 9.0 V to 12 V

44. Which of the following statements are true of the given type of DC-DC converter?

   I. The voltage source is the capacitor.
   II. The current source is the inductor.
   III. Forward energy transfer is used.
   IV. Flyback energy transfer is used.
   V. The output voltage is lower than the input voltage.
   VI. The output voltage is higher than the input voltage.

(A) buck converter: I, II, III, and VI
(B) buck converter: I, III, and VI only
(C) boost converter: I, II, and VI only
(D) boost converter: I, II, IV, and VI

45. An inductance of 5 mH has a current given by $i(t) = 1.5 \text{ A}(1 - e^{-1000t})$. What is most nearly the value of the voltage 100 μs into a transient?

(A) 0.0 V
(B) 6.8 V
(C) 7.5 V
(D) 13 V

46. A buck-boost single-quadrant converter operates with continuous current and has an input voltage of 28 V and a duty cycle of 80%. The output voltage is most nearly

(A) −110 V
(B) −22 V
(C) 28 V
(D) 110 V

47. The coaxial cable shown is used to communicate over a line with a 50 Ω resistance and a loss of 60 dB/100 ft at 5 GHz. The inner conductor has a radius, a, of 0.91 mm, and the outer conductor has an inner radius, b, of 4.95 mm. The dimensions of the cable are as shown along with the boundary conditions for the voltage.

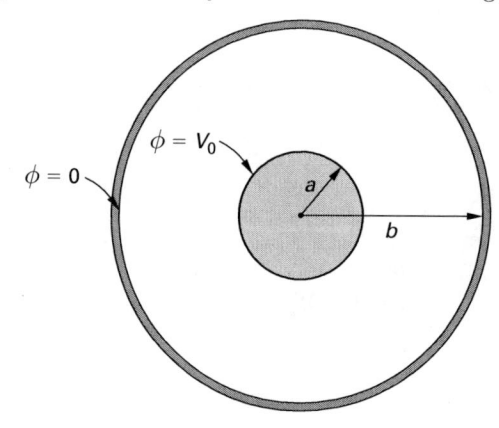

The DC potential applied, V_0, is 2.0 kV. At the cable-dielectric interface (radius = a), the transverse electric field is most nearly

(A) 1.3×10^3 V/m
(B) 1.4×10^3 V/m
(C) 1.3×10^6 V/m
(D) 1.4×10^6 V/m

48. The voltage in a circuit is represented by the complex number $3 + j5$ V. The current is represented by the complex number $4 - j10$ A. Which complex number represents the impedance?

(A) $0.10 + j0.43$ Ω
(B) $-0.47 + j0.20$ Ω
(C) $-0.33 + j0.43$ Ω
(D) $0.14 + j0.60$ Ω

49. Consider the following figure of a MOSFET.

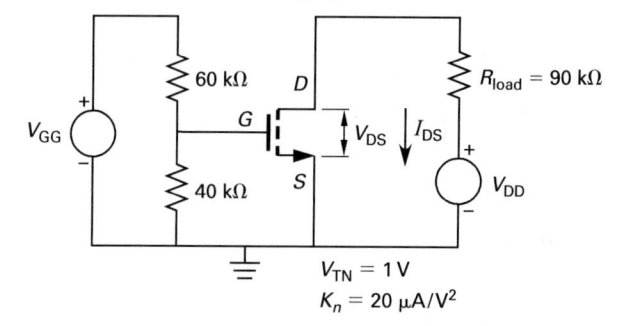

What is the gate-source voltage, V_{GS}, for the indicated configuration if the gate supply voltage, V_{GG}, is 10.0 V?

(A) 0 V

(B) 0.3 V

(C) 0.7 V

(D) 4.0 V

50. An optical transmission line is known to have an absolute index of refraction, n, of 2. Air generally approximates free space conditions for waves. What angle of incidence is required to obtain total internal reflection?

(A) $\pi/6$ rad

(B) $\pi/4$ rad

(C) $\pi/3$ rad

(D) $\pi/2$ rad

51. The maximum frequency that can be carried on a coaxial cable with an inner radius of 14.1 mm is most nearly

(A) 8.6 MHz

(B) 14 MHz

(C) 20 MHz

(D) 6.2 GHz

52. Consider the following circuit.

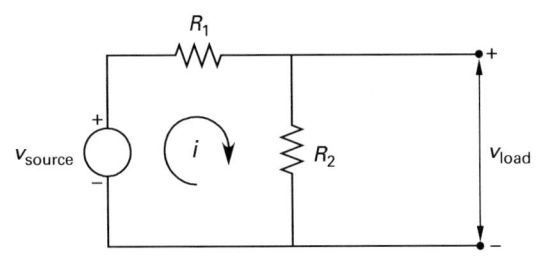

What is the open-loop transfer function for v_{load}?

(A) $v_{load} = iR_2$

(B) $i = \dfrac{v_{source}}{R_1 + R_2}$

(C) $v_{load} = \left(\dfrac{R_2}{R_1 + R_2}\right) v_{source}$

(D) $v_{load} = \left(\dfrac{v_{source} - v_{load}}{R_1}\right) R_2$

53. A transmitting antenna exhibits the properties shown.

output power, P_T	20,000 W
gain, G_T	5
cable losses, L_{cable}	1 dB

The path losses, taken from tabulated values of expected design conditions, are

scattering path loss, $L_{scatter}$	3 dB
free space path loss, L_{space}	25 dB
amplification path loss, L_{amp}	1 dB

The receiving antenna has a gain of 20 dB. The power at the receiving antenna is most nearly

(A) −4 dB

(B) 30 dB

(C) 40 dB

(D) 80 dB

54. An antenna has a physical aperture of 0.1 m² and a total efficiency of 70%. The antenna is meant to transmit at 2.4 GHz. The gain of the antenna is most nearly

(A) 0.8

(B) 5

(C) 6

(D) 60

55. During a stability study, the values of the source impedance and the load impedance are checked at steady-state conditions through the highest frequencies occurring during transients. At a given transient frequency of interest, the source and impedance values are calculated as

$$\mathbf{Z}_{source} = 0.01 + j1.2 \ \Omega$$
$$\mathbf{Z}_{load} = 0.5 - j0.8 \ \Omega$$

What is most nearly the magnitude of the source-to-load impedance ratio at this frequency, and does this indicate unconditional or conditional stability?

(A) 0.79, unconditionally stable

(B) 0.79, conditionally stable

(C) 1.3, unconditionally stable

(D) 1.3, conditionally stable

56. A given transistor circuit uses voltage-divider bias as shown.

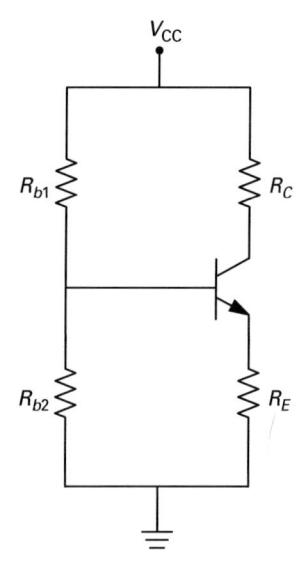

What is the primary purpose of the emitter resistor, R_E?

(A) It prevents thermal runaway.
(B) It increases sensitivity.
(C) It decreases output resistance.
(D) It sets the Q-point.

57. The closed-loop characteristic polynomial for a control system is given by $s^2 + 4s + 144$. What is most nearly the system damping ratio?

(A) 0.010
(B) 0.17
(C) 4.0
(D) 12

58. Which of the following statements referring to commercial AM radio broadcasts in the United States is FALSE?

(A) The carrier signal is suppressed.
(B) The modulated signal is frequency shifted.
(C) Both sidebands are utilized.
(D) The signal is generally DSB-LC.

59. A system has the following characteristic equation.

$$s^3 + 6s^2 + 3s + C = 0$$

Which of the following values of C ensures system stability?

(A) -6
(B) -5
(C) 11
(D) 30

60. A digital voltmeter using a 20 kΩ resistor is designed to measure over a frequency range of 0 Hz to 1000 Hz. What is most nearly the expected thermal agitation noise, or Johnson noise, at 25°C?

(A) 0.03 pV
(B) 0.3 pV
(C) 0.2 μV
(D) 0.6 μV

61. The magnetic field strength (in A/m) in a region of space is given by

$$\mathbf{H} = (5x + 3y)\mathbf{i} - (5y + 3z)\mathbf{j} + (5z - 3x)\mathbf{k}$$

The curl of the magnetic field strength determines the current density, $\mathbf{J}$ (in A/m²). What is $\mathbf{J}$ in this area of space?

(A) $3\mathbf{i} + 3\mathbf{j} - 3\mathbf{k}$ A/m²
(B) $0\mathbf{i} + 3\mathbf{j} - 3\mathbf{k}$ A/m²
(C) $-3\mathbf{i} + 3\mathbf{j} + 0\mathbf{k}$ A/m²
(D) $-3\mathbf{i} - 3\mathbf{j} + 6\mathbf{k}$ A/m²

62. The RF circuit shown uses a coupler that results in meter power, P_m, that is 3 dB less than the input coupler power, $P_{c,\text{in}}$.

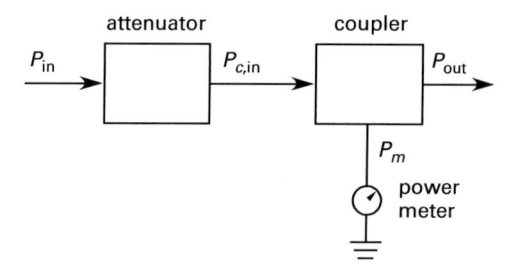

The attenuator results in a 10 dB loss over the frequency range of operation. The input power available is 30 mW. Determine the reading (in mW) on the power meter.

(A) 1.50 mW
(B) 1.77 mW
(C) 14.7 mW
(D) 17.0 mW

63. A digital voltmeter that is an integral part of a communications monitoring circuit is designed to operate under the following conditions.

$$T_{ambient} = 25°C$$
$$BW = 100 \text{ kHz}$$
$$P_{in} = 2 \text{ mW}$$

Determine the approximate thermal agitation noise at ambient conditions generated by a 10 kΩ resistor within the voltmeter.

(A) 1 pV
(B) 1 μV
(C) 4 μV
(D) 4 mV

64. The circuitry for an electronic voltmeter using a d'Arsonval display is shown.

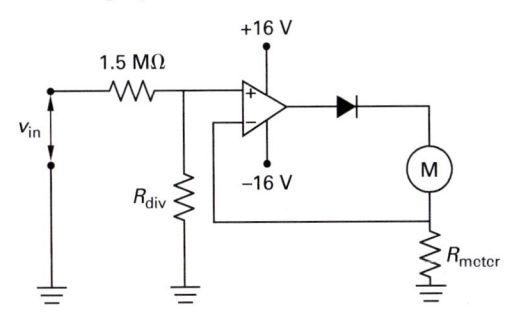

The full-scale meter current of 0.1 mA provides an rms reading of 200 V. Linear operation is ensured by maintaining the op amp input and output to within 3 V of the power-supply voltages. Determine the approximate value of divider resistance, R_{div}, necessary to ensure linear operation, assuming a sinusoidal input.

(A) 0.01 MΩ
(B) 0.07 MΩ
(C) 0.2 MΩ
(D) 0.6 MΩ

65. The input, v_{in}, to the computer circuitry shown is adjusted to the manufacturer's specification for obtaining a logic 1 output.

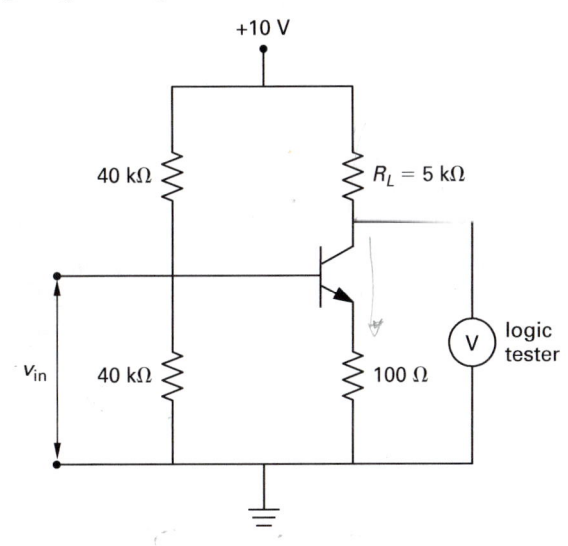

The data sheet provided with the circuitry indicates that the logic 1 condition occurs when the base-emitter junction and the base-collector junction are forward biased. An instrument called a logic tester is attached to the output terminals. The logic tester has two possible settings: positive logic and negative logic. Determine the approximate current through the load resistor and the position of the logic tester switch.

(A) 0 mA, negative
(B) 0 mA, positive
(C) 2 mA, negative
(D) 2 mA, positive

66. A portion of a control system circuit used as an on-off switch for the frequency of the incoming signal is shown.

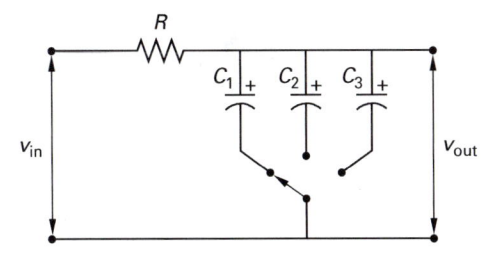

For the particular model of filter installed in the control system, the resistance is 50 kΩ. The values of the selectable capacitances are 0.01 μF, 0.05 μF, and 0.1 μF for C_1, C_2, and C_3, respectively. The "3 dB down point" is used as the cutoff. A control system is meant to respond to signals from the DC level to 60 Hz AC. Which capacitor meets this requirement?

(A) C_1
(B) C_2
(C) C_3
(D) none of the above

67. The circuit shown is used to measure the difference in voltage between inputs V_1 and V_2. The op amp properties of this circuit closely approximate those of an ideal op amp.

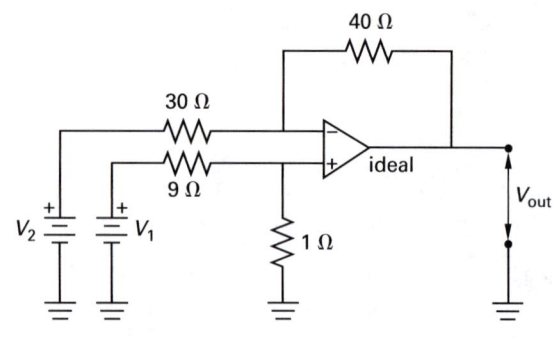

If $V_1 = 11$ V and $V_2 = 12$ V, what is the approximate output voltage?

(A) -16 V
(B) -13 V
(C) 1.0 V
(D) 15 V

68. The human body can be modeled as an approximately 500 Ω resistor. The voltage required to generate the standard let-go current is most nearly

(A) 5 V
(B) 10 V
(C) 30 V
(D) 50 V

69. A circuit is shown.

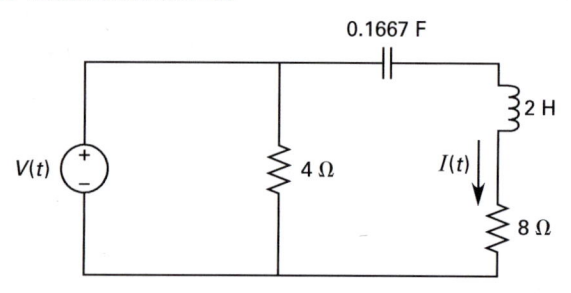

What is the transfer function, $I(s)/V(s)$, for the circuit?

(A) $\dfrac{2}{\dfrac{6}{s} + s + 8}$

(B) $\dfrac{4}{s^2 + 2s + 3}$

(C) $\dfrac{2s}{(s+2)(s+3)}$

(D) $\dfrac{2s}{(s+1)(s+3)}$

70. An AM radio station advertises their frequency as 680 kHz. What frequency is the local oscillator of an AM broadcast receiver set to in order to receive the advertised station?

(A) 225 kHz
(B) 455 kHz
(C) 680 kHz
(D) 1135 kHz

71. The FCC Rules, that is, the chapters associated with the requirements in the *Code of Federal Regulations*, limit a radio station to 5 kW power to avoid interference. The directivity gain of the antenna used is 1.4. What is most nearly the power density of the signal 5 km from the transmitting antenna?

(A) 22 μW/m²
(B) 0.10 mW/m²
(C) 20 W/m²
(D) 2.0 kW/m²

72. Communication systems use a variety of access methods to increase channel capacity while avoiding interference or collisions of signals. Match each abbreviation of the multiple-access technique to its appropriate description.

1. CSMA
2. TDMA
3. WDM

A. allows users to share the same channel by using different time slots

B. used in optical networks by sending multiple wavelengths

C. a type of media access control that ensures the absence of traffic before transmitting

(A) 1A, 2B, 3C
(B) 1B, 2A, 3C
(C) 1C, 2B, 3A
(D) 1C, 2A, 3B

73. Two techniques used in spread-spectrum modulation are frequency-hopping spread spectrum (FHSS) and direct-sequence spread spectrum (DSSS). Which of the following is NOT a property of the FHSS technique?

(A) expensive implementation
(B) allowable reuse of frequencies
(C) potential impact from interference
(D) high signal strength

74. Information is provided at 3000 Hz. What is the Nyquist rate to ideally sample this information signal and, in practice, should the sampling rate be higher or lower than this value?

(A) 1500 Hz, lower
(B) 3000 Hz, higher
(C) 6000 Hz, lower
(D) 6000 Hz, higher

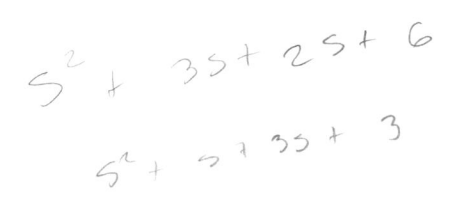

75. A component is expected to fail due to random causes only and does not experience extensive deterioration during its lifetime. The mean time to failure for the component is listed as 40,000 hr. After 3 yr of continuous operation, the component's reliability is most nearly

(A) 0.4
(B) 0.5
(C) 0.7
(D) 1

76. An electronic circuit is located in a panel that is powered from a 20 A circuit breaker. The required size of the safety grounding copper conductor is most nearly

(A) 8 AWG
(B) 10 AWG
(C) 12 AWG
(D) 14 AWG

77. The output of a two-port network is tested. With the output terminals shorted, the current is measured as 7 A∠80°. With the output terminals open, the voltage is measured as 28 V∠0°. The network's equivalent circuit with an rms voltage source in series with a complex impedance is most nearly

(A) 4.0 V∠0° and $0.69 - j3.9 \ \Omega$
(B) 4.0 V∠0° and $0.69 + j3.9 \ \Omega$
(C) 28 V∠0° and $0.69 - j3.9 \ \Omega$
(D) 28 V∠0° and $0.69 + j3.9 \ \Omega$

78. A uniform current flows through a wire in the direction shown.

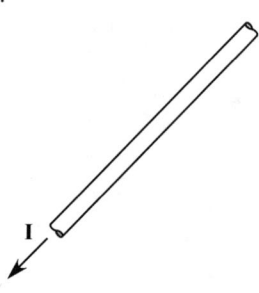

Which of the following represents the direction of the magnetic field?

(A)

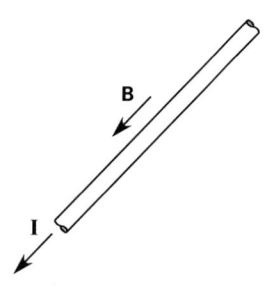

(B)

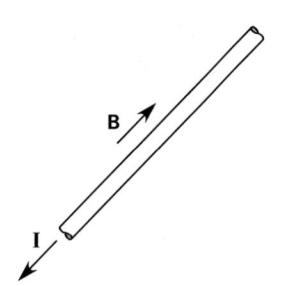

(C)

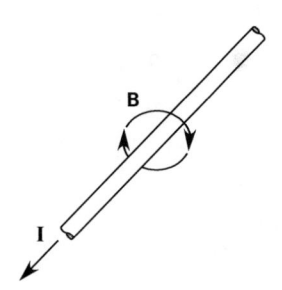

(D)

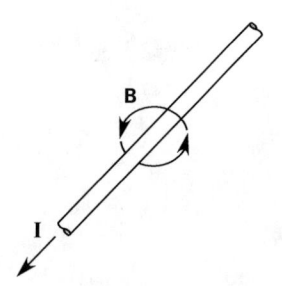

79. A digital circuit used to control the lights in a manufacturing center uses three inputs, A, B, and C, to determine the need for lighting. A Karnaugh map of the resulting logic analysis is shown.

C \ AB	00	01	11	10
0	1	0	d	1
1	1	d	1	d

A programmable logic array (PLA) will be used to realize the function. The unprogrammed PLA is shown.

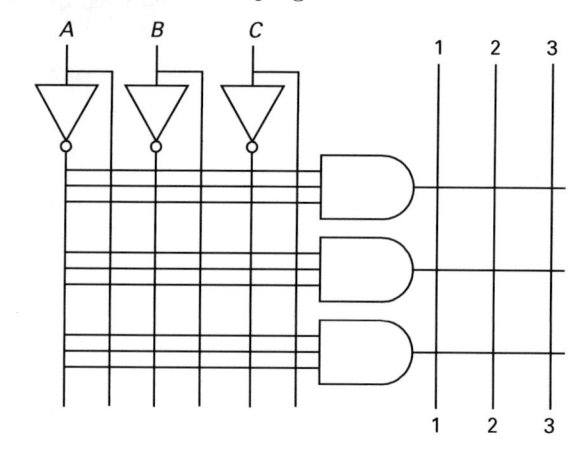

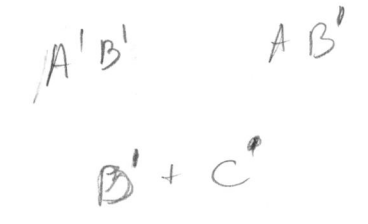

Which of the following PLAs realizes the minimum function, $F(A, B, C)$, on line 1?

(A)

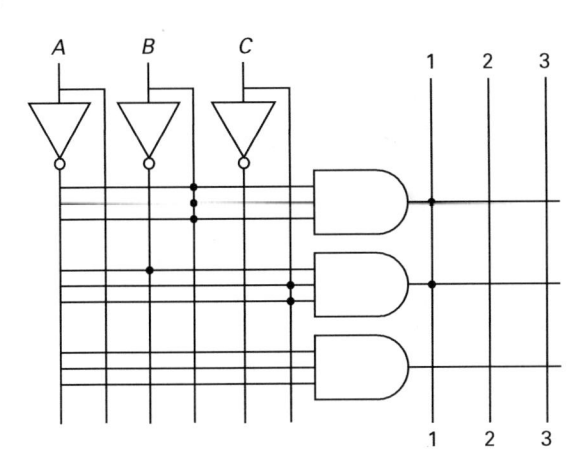

(B)

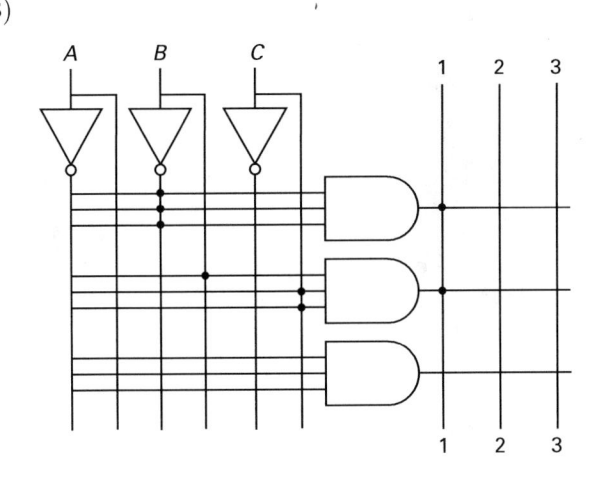

(C)

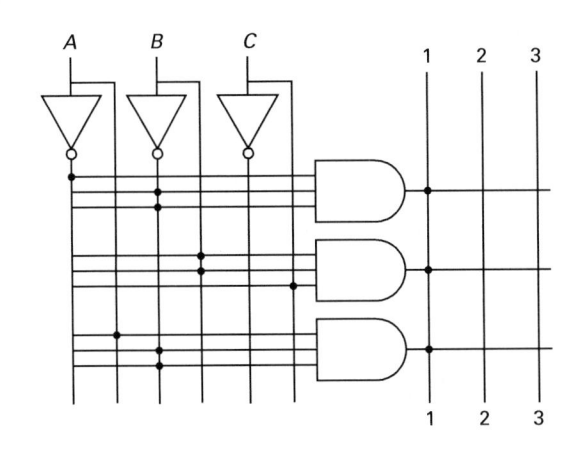

(D) ⭕

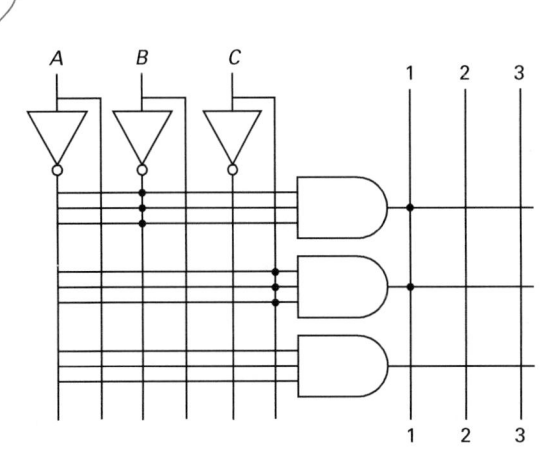

80. A manufacturing device is controlled by digital circuitry that uses a three-variable input. The truth table for the input and the required output, $F(X, Y, Z)$, is shown.

X	Y	Z	$F(X, Y, Z)$
0	0	0	0
0	0	1	1 $X'Y'Z$
0	1	0	0
0	1	1	1 $X'YZ$
1	0	0	0
1	0	1	1 $XY'Z$
1	1	0	0
1	1	1	1 XYZ

What is the canonical SOP form of the output function?

(A) ⭕ $\overline{X}\,\overline{Y}Z + \overline{X}YZ + X\overline{Y}Z + XYZ$

(B) $XY\overline{Z} + X\overline{Y}\,\overline{Z} + \overline{X}Y\overline{Z} + \overline{X}\,\overline{Y}\,\overline{Z}$

(C) $(\overline{X} + \overline{Y} + Z)(\overline{X} + Y + Z)(X + \overline{Y} + Z)$
$\times (X + Y + Z)$

(D) $(X + Y + \overline{Z})(X + \overline{Y} + \overline{Z})(\overline{X} + Y + \overline{Z})$
$\times (\overline{X} + \overline{Y} + \overline{Z})$

STOP!

DO NOT CONTINUE!

This concludes the Afternoon Session of the examination. If you finish early, check your work and make sure that you have followed all instructions. After checking your answers, you may turn in your examination booklet and answer sheet and leave the examination room. Once you leave, you will not be permitted to return to work or change your answers.

Answer Keys

$\frac{27}{40} = 67.5\%$

Morning Session Answer Key

1. C ✓
2. C ✗
3. A ✓
4. D ✗
5. A ✗
6. C ✓
7. C ✗
8. C ✗
9. A ✓
10. B ✓
11. D ✓
12. B ✓
13. D ✓
14. C ✓
15. A ✗
16. C ✓
17. D ✓
18. D ✓
19. C ✓
20. C ✗
21. A ✓
22. D ✓
23. A ✗
24. B ✓
25. C ✓
26. A ✗
27. D ✗
28. A ✓
29. C ✓
30. B ✓
31. B ✓
32. A ✓
33. D ✗
34. B ✓
35. D ✓
36. A ✓
37. A ✓
38. D ✓
39. C ✗
40. C ✗

Afternoon Session Answer Key

41. D ✓
42. A ✓
43. B ✗
44. D ✓
45. B ✓
46. A ✓
47. C ✗
48. C ✓
49. D ✓
50. A ✓
51. D ✓
52. C ✓
53. D ✗
54. D ✓
55. D ✓
56. A ✓
57. B ✓
58. A ✓
59. C ✓
60. D ✓
61. A ✓
62. A ✗
63. C ✓
64. B ✗
65. C ✗
66. C ✗
67. B ✗
68. A ✓
69. D ✗
70. D ✓
71. A ✗
72. D ✓
73. D ✗
74. C ✓
75. B ✗
76. C ✓
77. D ✗
78. D ✓
79. D ✓
80. A ✓

$\frac{27}{40} = 67.5\%$

Solutions
Morning Session Exam

1. The regulator has a dropout voltage of approximately 1.2 V. From the test data sheet given, for an input voltage, V_{in}, between 4.75 V and 10 V and an output current, I_{out}, between 0 mA and 800 mA, the expected output voltage is 3.3 V.

The answer is (C).

2. The operational amplifier has a reference voltage input, provided by the battery, on its positive terminal. Due to the various components near path 1, a rise in output voltage will cause the voltage on the negative terminal of the operational amplifier to rise as well. When this voltage rises above the reference voltage on the positive terminal, the operational amplifier biases the PNP transistor on its output to conduct. This sets up the path through 2 and 4 that "drains" the current from the input signal, thus lowering the voltage and protecting the circuit.

The answer is (C).

3. Closed-loop stability conditions require that at some frequency ω, when the open loop transfer function equals 1, the angle of the same must be less than 180°. So,

$$|G(j\omega)H(j\omega)| = 1$$
$$\angle G(j\omega)H(j\omega) < -180°$$

In this case, the open loop transfer function is

$$G(s)H(s) = K\left(\frac{80}{1+(30\text{ s})s}\right)e^{-t_d s}$$
$$= \frac{160e^{-t_d s}}{1+30s}$$

Setting the absolute value of the open loop transfer function allows calculation of the crossover frequency.

$$|G(s)H(s)| = \left|\frac{160e^{-t_d s}}{1+30s}\right| = 1$$

Because the absolute value is what is being determined, the following is true for the crossover frequency.

$$\left|\frac{(160)(1)}{\sqrt{(1)^2+(30\omega)^2}}\right| = 1$$

$$\frac{160}{\sqrt{1+900\omega^2}} = 1$$

$$\sqrt{1+900\omega^2} = 160$$

Solving for the crossover frequency gives

$$\omega = \sqrt{\frac{(160)^2-1}{900}} = 5.333 \text{ rad/s}$$

The angle at the crossover must be less than $-180°$ and can be found from

$$\angle G(j\omega_c)H(j\omega_c) = -t_d\omega_c - \arctan 30\omega_c > -\pi$$

The maximum delay time can be calculated from this restriction.

$$-t_d\omega_c - \arctan 30\omega_c > -\pi$$

$$t_d < \frac{\pi - \arctan 30\omega_c}{\omega_c}$$

$$= \frac{\pi - \arctan\left((30\text{ s})\left(5.333\ \frac{\text{rad}}{\text{s}}\right)\right)}{5.333\ \frac{\text{rad}}{\text{s}}}$$

$$= 0.2959 \text{ s} \quad (0.30 \text{ s})$$

The answer is (A).

4. The transform can be factored into two recognizable inverse Laplace transforms.

$$F(s) = \frac{2}{s^3(s-3)} = \left(\frac{2}{s^3}\right)\left(\frac{1}{s-3}\right) = F_1(s)F_2(s)$$

The inverse transform of $F_1(s)$ is t^2. The inverse transform of $F_2(s)$ is e^{3t}. Laplace transforms do not commute with ordinary multiplication. That is, $f_1(t)f_2(t) \neq f(t)$, even though $F_1(s)F_2(s) = F(s)$. Instead, convolution is used to determine $f(t)$. (This could also be determined using partial fractions.)

$$f(t) = \int_0^t f_1(\chi)f_2(t-\chi)\,d\chi$$
$$= \int_0^t \chi^2 e^{3(t-\chi)}\,d\chi$$
$$= e^{3t}\int_0^t \chi^2 e^{-3\chi}\,d\chi$$
$$= e^{3t}\left(\left.\frac{\chi^2 e^{-3\chi}}{-3}\right|_0^t - \frac{2}{-3}\int_0^t \chi e^{-3\chi}\,d\chi\right)$$
$$= e^{3t}\left(\frac{t^2 e^{-3t}}{-3} + \left(\frac{2}{3}\right)\left(\left.\frac{e^{-3\chi}}{(-3)^2}(-3\chi-1)\right|_0^t\right)\right)$$
$$= e^{3t}\left(\frac{t^2 e^{-3t}}{-3} + \left(\frac{2}{3}\right)\left(\begin{array}{l}\frac{e^{-3t}}{9}(-3t-1)\\-\frac{e^{(-3)(0)}}{9}((-3)(0)-1)\end{array}\right)\right)$$
$$= e^{3t}\left(\frac{t^2 e^{-3t}}{-3} + \left(\frac{2}{3}\right)\left(\frac{e^{-3t}}{9}(-3t-1)+\frac{1}{9}\right)\right)$$
$$= e^{3t}\left(\frac{t^2 e^{-3t}}{-3} + \left(\frac{2e^{-3t}}{27}\right)(-3t-1)+\frac{2}{27}\right)$$
$$= e^{3t}\left(\frac{t^2 e^{-3t}}{-3} - \frac{6te^{-3t}}{27} - \frac{2e^{-3t}}{27}+\frac{2}{27}\right)$$
$$= -\frac{t^2}{3} - \frac{2}{9}t - \frac{2}{27} + \frac{2}{27}e^{3t}$$
$$= \frac{2}{27}e^{3t} - \frac{1}{3}t^2 - \frac{2}{9}t - \frac{2}{27}$$

The answer is (D).

5. Traditionally, problems such as this are worked in the time domain. The use of the phasor concept often simplifies this type of problem and will be used here for edification.

The "ideal conditions" apply to the op amp and also to the discrete elements; that is, the op amp is ideal and the feedback capacitor exhibits no leakage. (Such leakage would make the circuit a "leaky integrator," and the resulting solution, though similar, involves a phase shift.)

Consider node A to be at the negative input to the op amp.

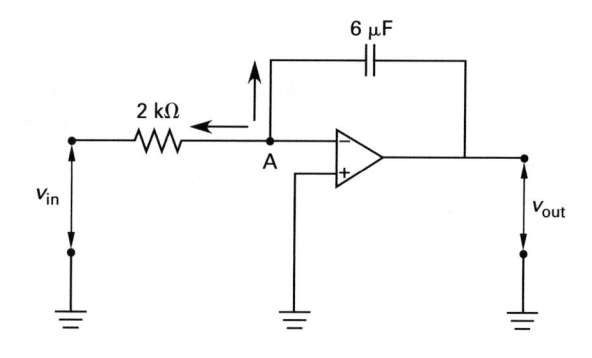

Apply Kirchhoff's current law at node A.

$$\frac{0\text{ V} - \mathbf{V}_{in}}{Z_R} + \frac{0\text{ V} - \mathbf{V}_{out}}{Z_C} = 0\text{ A}$$

Solve for the output voltage.

$$\mathbf{V}_{out} = \left(-\frac{\mathbf{Z}_C}{\mathbf{Z}_R}\right)\mathbf{V}_{in}$$

The term $-\mathbf{Z}_C/\mathbf{Z}_R$ is the gain factor, which includes a negative sign to indicate the inversion of the signal. Substitute the given information, in phasor form, to determine an expression for the output voltage.

$$\mathbf{V}_{out} = \left(-\frac{\mathbf{Z}_C}{\mathbf{Z}_R}\right)\mathbf{V}_{in} = \left(-\frac{\frac{1}{j\omega C}}{R}\right)\mathbf{V}_{in}$$
$$= \left(-\frac{1}{j\omega CR}\right)\mathbf{V}_{in}$$
$$= \left(-\frac{1}{(1\angle 90°)(377\text{ Hz})}\times(6\times 10^{-6}\text{ F})(2\times 10^3\text{ }\Omega)\right)(1\text{ V}\angle 0°)$$
$$= 0.221\text{ V}\angle 90°$$

Change this result into the time domain.

$$\mathbf{V}_{out} = 0.221 \text{ V} \angle 90°$$
$$v_{out} = (0.221 \text{ V})\sin(377t + 90°) + K$$
$$= (0.221 \text{ V})\cos 377t + K$$

K is an arbitrary constant.

Apply the boundary condition $v_{out} = 0$ V at $t = 0$ s to determine the value of K.

$$v_{out}(0) = (0.221 \text{ V})\cos 377t + K$$
$$= 0 \text{ V}$$
$$K = (-0.221 \text{ V})\cos\bigl((377)(0 \text{ V})\bigr)$$
$$= -0.221 \text{ V}$$

Substitute the value of the constant in the expression for the output voltage. The final solution is

$$v_{out} = (0.221 \text{ V})\cos 377t - 0.221 \text{ V}$$
$$= (0.221 \text{ V})(\cos 377t - 1)$$

The answer is (A).

6. Using the node voltage method, the circuit currents are as shown.

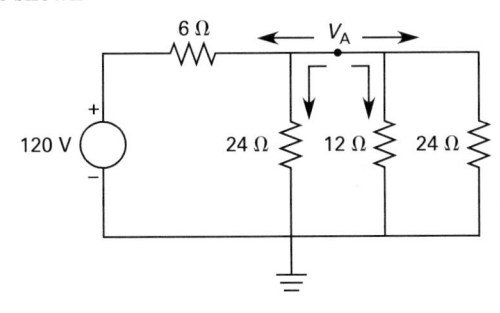

Apply Kirchhoff's current law at node A.

$$\frac{V_A - 120 \text{ V}}{6 \text{ }\Omega} + \frac{V_A}{24 \text{ }\Omega} + \frac{V_A}{12 \text{ }\Omega} + \frac{V_A}{24 \text{ }\Omega} = 0 \text{ A}$$
$$\frac{V_A}{6 \text{ }\Omega} - \frac{120 \text{ V}}{6 \text{ }\Omega} + \frac{V_A}{24 \text{ }\Omega} + \frac{V_A}{12 \text{ }\Omega} + \frac{V_A}{24 \text{ }\Omega} = 0 \text{ A}$$
$$V_A\left(\frac{1}{6 \text{ }\Omega} + \frac{1}{24 \text{ }\Omega} + \frac{1}{12 \text{ }\Omega} + \frac{1}{24 \text{ }\Omega}\right) = \frac{120 \text{ V}}{6 \text{ }\Omega}$$
$$V_A = 60 \text{ V}$$

The answer is (C).

7. The possible op amp circuits, with their names, are repeated for convenience.

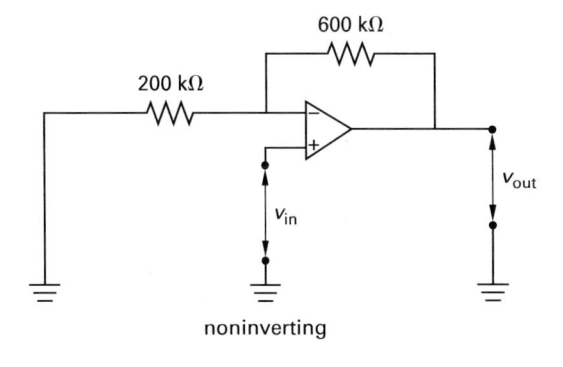

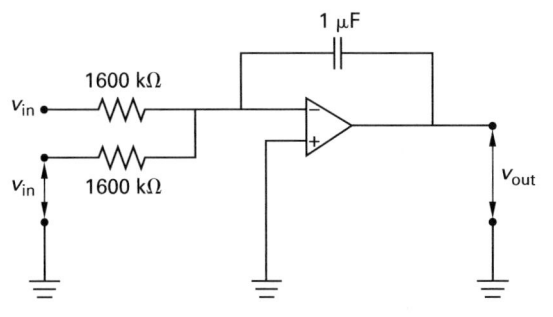

noninverting

integrator summer

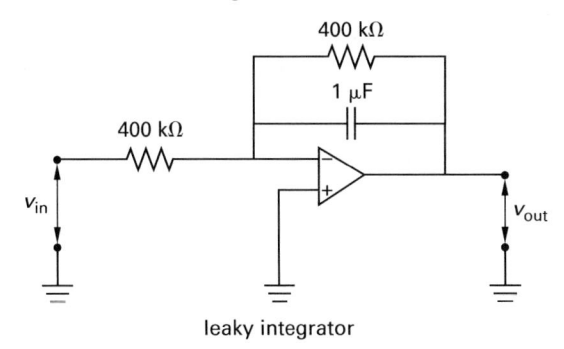

leaky integrator

voltage follower

Assuming ideal op amp properties, only one of the possibilities given has an output resistance of 0 Ω—the circuit in option D. Ensure this is the correct selection by checking the gain of this voltage follower, using the concept of a voltage divider.

$$v_{out} = v^+ = v_{in}\left(\frac{600 \text{ k}\Omega}{800 \text{ k}\Omega}\right)$$
$$= 0.75 v_{in}$$

The gain is 0.75.

The input resistance is the series combination of the 200 kΩ and 600 kΩ resistor, or 800 kΩ as required.

The answer is (D).

8. The internal resistance of the inductor affects the steady-state time but not the resonant frequency. Accounting for the resistance gives a series RLC circuit as shown.

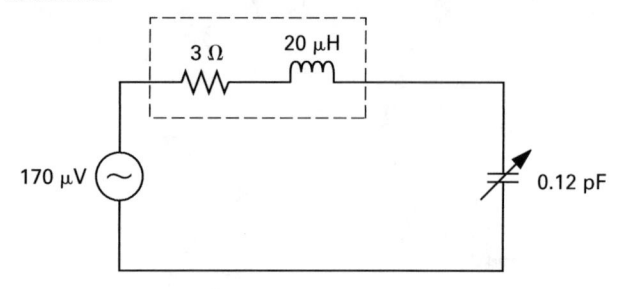

The resonant frequency of a series RLC circuit is

$$f_0 = \frac{1}{2\pi\sqrt{LC}}$$

$$= \frac{1}{2\pi\sqrt{(20 \times 10^{-6} \text{ H})(0.12 \times 10^{-12} \text{ F})}}$$

$$= 102.7 \times 10^6 \text{ Hz} \quad (100 \text{ MHz})$$

The answer is (C).

9. The wavelength of a given electromagnetic wave in free space is

$$\lambda = \frac{c}{f}$$

c represents the speed of light. The velocity factor, k, is the ratio of the speed of the electromagnetic wave in a given transmission line or cable (for example v) to the speed in free space.

$$k = \frac{\text{v}}{c}$$

Combining the two equations gives the wavelength of an electromagnetic wave in a cable with velocity factor k.

$$\lambda = \frac{kc}{f}$$

Substitute the given information.

$$\lambda = \frac{(0.6)\left(3 \times 10^8 \ \frac{\text{m}}{\text{s}}\right)}{1 \times 10^9 \text{ Hz}}$$

$$= 0.18 \text{ m} \quad (0.2 \text{ m})$$

The answer is (A).

10. The period of a given waveform is

$$T = \frac{1}{f}$$

The frequency is

$$\omega = 2\pi f$$

$$f = \frac{\omega}{2\pi} = \frac{2513 \text{ Hz}}{2\pi}$$

$$= 400 \text{ Hz}$$

Substitute the calculated value to determine the period.

$$T = \frac{1}{f} = \frac{1}{400 \text{ Hz}}$$

$$= 2.5 \times 10^{-3} \text{ s} \quad (2.5 \text{ ms})$$

The answer is (B).

11. This solution involves the complex frequency $\mathbf{s} = \sigma + j\omega$. Use only the real portion of $e^{jxt} = \cos xt + j\sin xt$, given a constant A and factor $e^{\sigma t}$.

$$Ae^{\sigma t}e^{j(\omega t + \phi)} = Ae^{\sigma t}e^{j\omega t}e^{j\phi}$$

$$= Ae^{j\omega t + \sigma t}e^{j\phi}$$

$$Ae^{(\sigma + j\omega)t}e^{j\phi} = Ae^{\mathbf{s}t}e^{j\phi}$$

$$= Ae^{j\phi}e^{\mathbf{s}t}$$

$$\text{Re}\left(Ae^{\sigma t}e^{j(\omega t + \phi)}\right) = Ae^{\sigma t}\cos(\omega t + \phi)$$

Using the relationships in the listed equations, consider $3e^{-2t}\cos(377t - 30°)$. The constant term A is 3 and the term ϕ is 30°, neither of which is included in $\mathbf{s}$, since these terms represent values that do not change with time. The term σ, which is also called the neper frequency, is -2 Hz. The angular frequency ω is 377 Hz. The complex frequency, which is used to represent the time-varying portions of the signal, is

$$\mathbf{s} = -2 + j377$$

The complex frequency $\mathbf{s}$ is usually written as s when used in the exponent, e^{st}.

The answer is (D).

12. To evaluate the purpose of the circuit, the ratio V_{out}/V_{in} is needed. To obtain this ratio, note that for ideal operational amplifier characteristics, V_+ and V_- are equal. Use voltage divider concepts to find expressions for V_+ and V_-. For the positive terminal input,

$$V_+ = \left(\frac{\frac{1}{sC}}{R + \frac{1}{sC}}\right)V_{in} = \left(\frac{\frac{1}{sC}}{\frac{sRC}{sC} + \frac{1}{sC}}\right)V_{in}$$

$$= \left(\frac{1}{1 + sRC}\right)V_{in}$$

For the negative terminal input,

$$V_- = \left(\frac{R_G}{R_G + R_F}\right)V_{out} = \left(\frac{1}{1 + \frac{R_F}{R_G}}\right)V_{out}$$

Equate the two expressions, and solve for the ratio V_{out}/V_{in}.

$$\left(\frac{1}{1 + sRC}\right)V_{in} = \left(\frac{1}{1 + \frac{R_F}{R_G}}\right)V_{out}$$

$$\frac{V_{out}}{V_{in}} = \left(1 + \frac{R_F}{R_G}\right)\left(\frac{1}{1 + sRC}\right)$$

To determine the type of filter, determine the value of V_{out}/V_{in} at $\omega = 0$ and $\omega = \infty$. The complex frequency, s, contains the frequency term $j\omega$.

At $\omega = 0$, V_{out}/V_{in} is determined by the resistors, so that some value of the input is passed to the output.

$$\frac{V_{out}}{V_{in}} = \left(1 + \frac{R_F}{R_G}\right)\left(\frac{1}{1 + sRC}\right)$$

$$= \left(1 + \frac{R_F}{R_G}\right)\left(\frac{1}{1 + 0}\right)$$

$$= 1 + \frac{R_F}{R_G}$$

At $\omega = \infty$, V_{out}/V_{in} is infinitely small, making the output zero.

$$\frac{V_{out}}{V_{in}} = \left(1 + \frac{R_F}{R_G}\right)\left(\frac{1}{1 + \infty}\right)$$

$$= \left(1 + \frac{R_F}{R_G}\right)(0)$$

$$= 0$$

The circuitry passes low frequency but inhibits high frequency, so this is a low-pass filter.

The answer is (B).

13. The op amps are used to compare the input voltage on the positive (+) terminal to the reference voltage on the negative (−) terminal. The output of the op amp drives to the positive supply voltage (+5 V) if the positive input terminal voltage magnitude is higher, and to the negative supply voltage (−5 V) if the negative input terminal voltage magnitude is higher.

The negative input terminal voltage is set by the resistor network as shown.

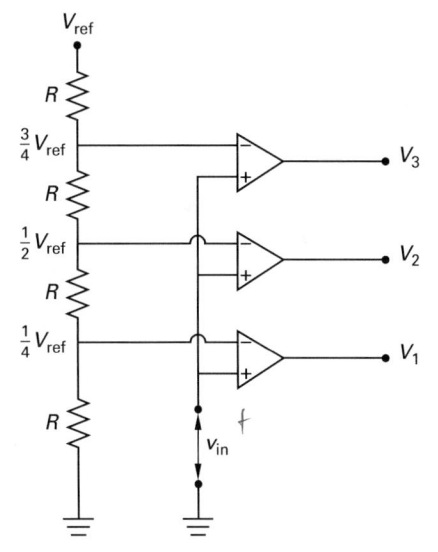

The voltage at $\frac{3}{4}V_{ref}$ is 0.75 V. The input voltage is 0.76 V. Therefore, the voltage on the positive input terminal is greater than the voltage on the negative input terminal on all three op amps. The output of all the op amps will be at the positive power-supply voltage level of +5 V, or HIGH (H, H, H).

The answer is (D).

14. Assume the op amp can be accurately approximated as ideal, which is normally the case. The directions used to apply Kirchhoff's current law (KCL) at the negative input terminal are

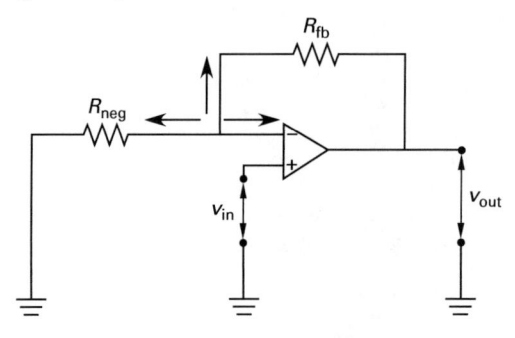

Apply KCL at the negative input terminal.

$$\frac{v_{in} - 0\text{ V}}{R_{neg}} + \frac{v_{in} - v_{out}}{R_{fb}} + 0\text{ A} = 0\text{ A}$$

Rearrange the equation to solve for the output voltage expression.

$$\frac{v_{out}}{R_{fb}} = \frac{v_{in}}{R_{neg}} + \frac{v_{in}}{R_{fb}}$$

$$v_{out} = v_{in}\left(\frac{R_{fb}}{R_{neg}} + 1\right)$$

$$= v_{in}\left(1 + \frac{R_{fb}}{R_{neg}}\right)$$

The output is not inverted, and the gain is greater than 1 for this configuration.

The answer is (C).

15. Assume the op amp is ideal.

$$v^- = v^+$$

The voltages at the input terminals are

$$v^- = v_{out}$$
$$v^+ = v_{in}$$

Given the observation that the input terminal voltages are equal,

$$v_{out} = v_{in}$$

The loading effect of the load resistor on the source is eliminated, hence the name "buffer" for the op amp configuration. (This is also called a voltage follower, since the output voltage follows the input voltage.)

The answer is (A).

16. The modulated signal is given by

$$m(t) = (1 + 3.6\sin 25t)(4\sin 100t)$$
$$= 4\sin 100t + 14.4\sin 25t \sin 100t$$

Half the power is at $+\omega_c$ and half is at $-\omega_c$, so the modulated signal can be rewritten using trigonometric relationships as

$$m(t) = 4\sin 100t + 7.2\,(\cos 75t - \cos 125t)$$

7.2 represents the peak voltage of the signal. The power can be calculated as follows. The power is normalized to 1 Ω to allow use of the information for other circuits' actual resistance.

$$P = \frac{V_{rms}^2}{R} = \frac{\left(\frac{V_{peak}}{\sqrt{2}}\right)^2}{R} = \frac{\left(\frac{7.2\text{ V}}{\sqrt{2}}\right)^2}{1\text{ Ω}}$$
$$= 25.92\text{ W} \quad (26\text{ W})$$

The answer is (C).

17. Use the mesh current method with loop currents as shown.

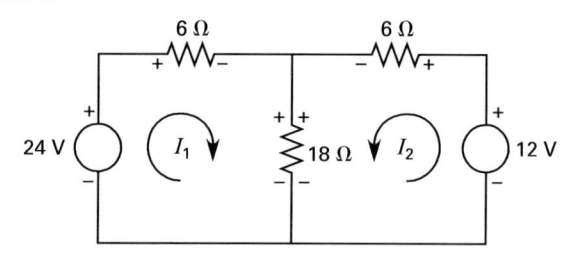

Apply Kirchhoff's voltage law (KVL) in loop 1.

$$24\text{ V} - I_1(6\text{ Ω}) - (I_1 + I_2)(18\text{ Ω}) = 0\text{ V}$$

Rearrange.

$$I_1 + \tfrac{3}{4}I_2 = 1\text{ A} \quad\quad [\text{I}]$$

Apply KVL in loop 2.

$$12\text{ V} - I_2(6\text{ Ω}) - (I_1 + I_2)(18\text{ Ω}) = 0\text{ V}$$

Rearrange.

$$I_1 + \tfrac{4}{3}I_2 = 2/3\text{ A} \quad\quad [\text{II}]$$

Solve Eqs. I and II simultaneously.

$$I_1 = \frac{10}{7} \text{ A}$$
$$I_2 = -\frac{4}{7} \text{ A}$$

The total current in the 18 Ω resistor is the sum of the loop currents.

$$I_t = I_1 + I_2 = \frac{10}{7} \text{ A} + \left(-\frac{4}{7} \text{ A}\right)$$
$$= 6/7 \text{ A}$$

The answer is (D).

18. Use the equation for the noise figure, NF, and solve for the output SNR.

$$\text{NF} = 10 \log F = 10 \log \frac{\text{SNR}_{\text{in}}}{\text{SNR}_{\text{out}}}$$
$$= \text{SNR}_{\text{in,dB}} - \text{SNR}_{\text{out,dB}}$$
$$\text{SNR}_{\text{out,dB}} = \text{SNR}_{\text{in,dB}} - \text{NF}$$
$$= 2.0 \text{ dB} - 1.5 \text{ dB}$$
$$= 0.5 \text{ dB}$$

The answer is (D).

19. The standard noise temperature is 290K. The noise factor for the chip is

$$F = 1 + \frac{T_{\text{noise}}}{T_0}$$
$$= 1 + \frac{300\text{K}}{290\text{K}}$$
$$= 2.03 \quad (2)$$

The answer is (D).

20. The charge on the capacitor is found using the following fundamental relationship.

$$Q = CV$$
$$= (15 \times 10^{-6} \text{ F})(90 \text{ V})$$
$$= 1.35 \times 10^{-3} \text{ C} \quad (1.4 \text{ mC})$$

The answer is (C).

21. The Norton equivalent circuit is defined by the open-circuit voltage and the short-circuit current as shown. (The resistors have been identified with subscripts for clarification.)

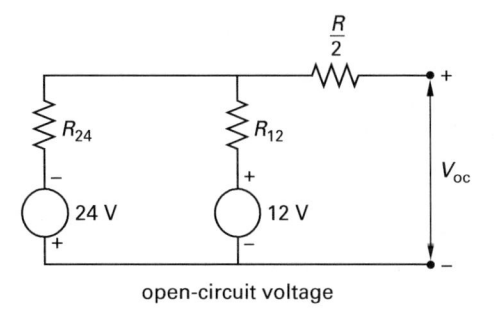

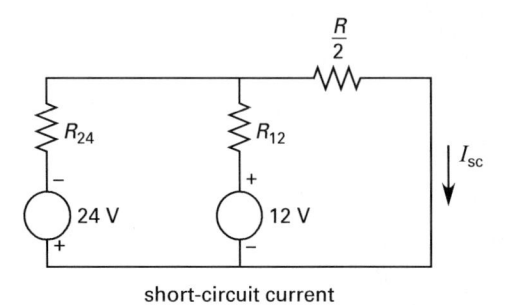

open-circuit voltage

short-circuit current

The solution can be determined using a number of different methods. Using superposition, the current from the 12 V source is

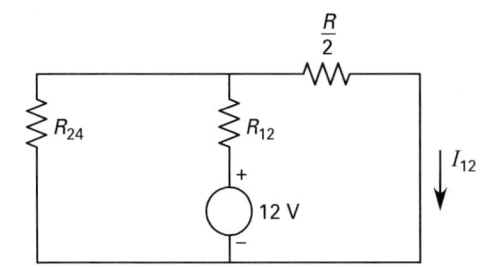

Using the concept of a current divider, I_{12} is

$$I_{12} = \left(\frac{12 \text{ V}}{R_{12} + \dfrac{R_{24}\left(\dfrac{R}{2}\right)}{R_{24} + \dfrac{R}{2}}}\right)\left(\frac{R_{24}}{R_{24} + \dfrac{R}{2}}\right)$$

$$= \left(\frac{12 \text{ V}}{R + \dfrac{R}{3}}\right)\left(\frac{R}{\frac{3}{2}R}\right)$$

$$= \frac{6 \text{ V}}{R}$$

Using superposition, the current from the 24 V source is

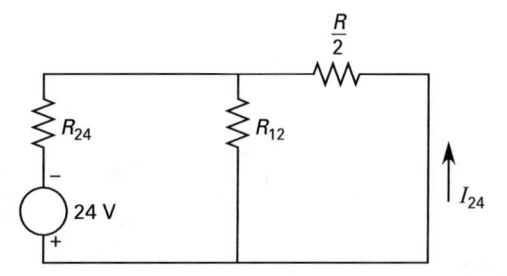

Using the concept of a current divider, I_{24} is

$$I_{24} = \left(\frac{24\text{ V}}{R_{24} + \dfrac{R_{12}\left(\dfrac{R}{2}\right)}{R_{12} + \dfrac{R}{2}}}\right)\left(\frac{R_{12}}{R_{12} + \dfrac{R}{2}}\right)$$

$$= \left(\frac{24\text{ V}}{R + \dfrac{R}{3}}\right)\left(\frac{R}{\dfrac{3}{2}R}\right)$$

$$= \frac{12\text{ V}}{R}$$

The short-circuit current, as defined, is

$$I_{SC} = I_{12} - I_{24} = \frac{6\text{ V}}{R} - \frac{12\text{ V}}{R} = -\frac{6\text{ V}}{R}$$

$$= -\frac{6\text{ V}}{2\text{ }\Omega}$$

$$= -3\text{ A}$$

The resulting circuit is

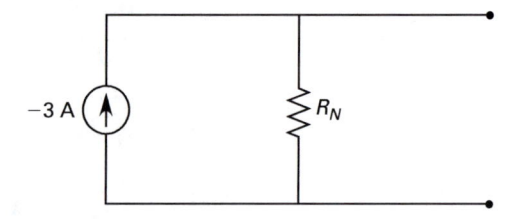

The magnitude of the current is 3.0 A.

The answer is (A).

22. The traffic capacity in dimensionless erlangs, E, is the product of the call rate, λ, and the call-holding time (average call time), h. For the traffic capacity to be dimensionless, the call rate and the call-holding time must be measured with the same units of time.

$$E = \lambda h$$

$$= \left(\frac{90{,}000\ \dfrac{1}{\text{hr}}}{60\ \dfrac{\text{min}}{\text{hr}}}\right)(3\text{ min})$$

$$= 4500\text{ erlangs}$$

The answer is (C).

23. The fundamental properties of an inductor can be modeled as a short circuit for low frequencies (DC) and an open circuit for high frequencies. The fundamental properties of a capacitor can be modeled as an open circuit for low frequencies and a short circuit for high frequencies.

For both the high- and low-frequency cases, one of the two circuit elements in the input circuitry would act as an open circuit, preventing any of the input voltage from reaching the load resistor, R. That is, the normalized input impedance is high at both low- and high-frequency levels. The only frequency response that meets these conditions is option A, which is the response of a circuit displaying series resonance.

The answer is (A).

24. Transforming the individual circuit elements into the s-domain and accounting for the initial capacitor voltage results in the circuit shown.

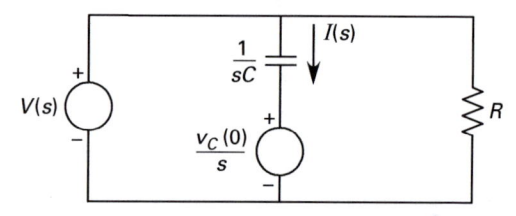

Use Ohm's law across the capacitor, rearranging for the current.

$$I(s) = \frac{V(s) - \dfrac{v_C(0)}{s}}{\dfrac{1}{sC}}$$

$$= sCV(s) - Cv_C(0)$$

$$= C\bigl(sV(s) - v_C(0)\bigr)$$

The answer is (B).

25. Use the equation for the energy of the light emitted, and solve for the frequency, ν.

$$E = \frac{hc}{\lambda} = hf = h\nu$$

$$\nu = \frac{E}{h} = \frac{(1.1 \text{ eV})\left(1.6022 \times 10^{-19} \frac{\text{J}}{\text{eV}}\right)}{6.6256 \times 10^{-34} \text{ J·s}}$$

$$= 2.660 \times 10^{14} \text{ Hz} \quad (270 \text{ THz})$$

In optical engineering, the Greek letter nu, ν, is used for frequency instead of f.

The answer is (D).

26. Timer setup A is for bistable operation, which is represented by output 1. The only possible correct combination is option A.

Timer setup B is for monostable operation. Output 2 represents astable operation.

The answer is (A).

27. From the given information, the gains of each of the functional parts of the phase-locked loop (PLL) are as follows.

$$K_{\text{PD}} = 0.15 \text{ V/rad}$$
$$K_{\text{LF}} = 4$$
$$K_{\text{VCO}} = 2.2 \times 10^6 \text{ Hz/V}$$

The open loop gain is

$$K_{\text{PLL}} = K_{\text{PD}} K_{\text{LF}} K_{\text{VCO}}$$
$$= \left(0.15 \frac{\text{V}}{\text{rad}}\right)(4)\left(2.2 \times 10^6 \frac{\text{Hz}}{\text{V}}\right)$$
$$= 1.32 \times 10^6 \text{ Hz/rad}$$

The equation for open loop gain can be used to calculate the change in VCO output frequency when the input frequency changes by 0.2 rad.

$$\frac{d\omega_{\text{out}}}{d\Delta\phi} = K_{\text{PLL}}$$
$$d\omega_{\text{out}} = (d\Delta\phi) K_{\text{PLL}}$$
$$= (0.2 \text{ rad})\left(1.32 \times 10^6 \frac{\text{Hz}}{\text{rad}}\right)$$
$$= 264{,}000 \text{ Hz} \quad (0.26 \text{ MHz})$$

This change in VCO frequency is then fed back to the input to drive the phase difference back to zero.

Though a radian is a dimensionless ratio, in problems such as this it is important to maintain unit consistency between the cycles per second in the unit hertz and the use of radians elsewhere. Equate the two with

$$1 \text{ Hz} = 1 \frac{\text{cycle}}{\text{s}}$$
$$= \left(1 \frac{\text{cycle}}{\text{s}}\right)\left(2\pi \frac{\text{rad}}{\text{cycle}}\right)$$
$$= 2\pi \text{ rad/s}$$

The answer is (B).

28. Since the field direction does not vary, the magnitude of the displacement current density is

$$J_D = \frac{\partial D}{\partial t} = \frac{\partial(\epsilon_0 \epsilon_r E)}{\partial t}$$
$$= \left(175 \frac{\text{V}}{\text{m}}\right) \epsilon_0 \epsilon_r \frac{\partial \sin(10)^9 t}{\partial t}$$
$$= \left(175 \frac{\text{V}}{\text{m}}\right)\left(8.854 \times 10^{-12} \frac{\text{F}}{\text{m}}\right)$$
$$\times (1)(10^9 \text{ Hz}) \cos 10^9 t$$
$$= \left(1.55 \frac{\text{A}}{\text{m}^2}\right) \cos 10^9 t$$

The answer is (A).

29. When data is not being transmitted, the transmission line of a universal asynchronous receiver/transmitter (UART) is normally held at a high voltage level, not a low one. The transition from high voltage to low voltage starts the transfer of data, with the start bit queueing the UART to the beginning of an information packet. Option D is correct.

A UART transmits data asynchronously, so no controlling clock is required. Data packets are limited but can be no more than 9 bits, or 8 bits if a parity bit is used.

The answer is (D).

30. The overall gain, K, without feedback is

$$K = K_1 K_2 K_3 = (-50)(-75)(-80) = -300\,000$$

The fraction of the output signal appearing at the summing point, h, depends on the resistances of the voltage divider.

$$h = \frac{R_{25}}{R_{25} + R_{1000}} = \frac{25 \text{ }\Omega}{25 \text{ }\Omega + 1000 \text{ }\Omega} = 0.024$$

The feedback is summed (i.e., adds to the input), so the positive feedback formula for gain is used. But since the gain is itself negative ($K < 0$), the overall feedback is negative.

$$K_{\text{loop}} = \frac{K}{1 - Kh} = \frac{-300\,000}{1 - (-300\,000)(0.024)}$$
$$= -41.66 \quad (-40)$$

The answer is (B).

31. The denominator of the transfer function exhibits the general form of the characteristic equation of a higher-order circuit.

$$as^2 + bs + c$$

For control systems, the characteristic equation is

$$s^2 + 2\zeta\omega_n s + \omega_n^2$$

For resonance conditions, this equation takes the form

$$s^2 + \left(\frac{\omega_0}{Q}\right)s + \omega_0^2$$

Compare the denominator of the given transfer function with the resonant form of the characteristic equation. Note that the magnitude of the resonant frequency squared is 10^6, making the resonant frequency magnitude 10^3.

$$20\,\frac{\text{rad}}{\text{s}} = \frac{\omega_0}{Q}$$

$$Q = \frac{\omega_0}{20\,\frac{\text{rad}}{\text{s}}} = \frac{10^3\,\frac{\text{rad}}{\text{s}}}{20\,\frac{\text{rad}}{\text{s}}}$$
$$= 50$$

The answer is (B).

32. The "information unit" is one bit, so one baud and one bit are identical, and 9600 baud is equivalent to 9600 bits/s. The period is

$$T = \frac{1}{R_{\text{baud}}}$$
$$= \frac{1}{9600\,\frac{\text{bits}}{\text{s}}}$$
$$= 0.0001042\,\text{s} \quad (100\,\mu\text{s})$$

The answer is (A).

33. The forcing function is a step of height 2 at $t = 0$ s. The Laplace transform of the unit step is $1/s$. Therefore, $F(s) = 2/s$. The response function is

$$R(s) = F(s)T(s) = \left(\frac{2}{s}\right)\left(\frac{6}{s+2}\right)$$
$$= \frac{12}{s(s+2)}$$

Using partial fractions, the response function can be represented as

$$R(s) = \frac{12}{s(s+2)}$$
$$= \frac{6}{s} - \frac{6}{s+2}$$

The time-based response function is found using the inverse Laplace transform of the response function.

$$r(t) = \mathcal{L}^{-1}\left(\frac{6}{s} - \frac{6}{s+2}\right)$$
$$= 6 - 6e^{-2t}$$

The answer is (D).

34. The final value of the time-based $r(t)$ is found from the frequency based $R(s)$ using the final value theorem, which states

$$\lim_{t\to\infty} r(t) = \lim_{s\to 0}(sR(s))$$

Multiply $R(s)$ by s and apply the limit.

$$r(\infty) = \lim_{s\to 0}(sR(s)) = \lim_{s\to 0}\frac{s}{s+1}$$
$$= \lim_{s\to 0}\frac{1}{s+1}$$
$$= 1$$

The answer is (B).

35. All the statements are true with the exception of option D. The bucket brigade is a method for approximating the dead time in a given controller.

The answer is (D).

36. The numerator is zero when $s = -4$. This is the only zero in the transfer function. (Thus, option A should be the correct answer.)

The denominator is zero when $s = -2$ and $s = -1 \pm j$. The later zero can be seen by factoring the quadratic in the denominator. These three values are the poles of the transfer function.

The answer is (A).

37. The phase margin is the number of degrees the phase angle is above $-180°$ at the gain crossover point (i.e., where the logarithmic gain is 0 dB or the actual gain is 1). This is point B.

The answer is (B).

38. The reflection coefficient of the transmission line is

$$\Gamma_L = \frac{Z_L - Z_0}{Z_L + Z_0} = \frac{120\ \Omega - 60\ \Omega}{120\ \Omega + 60\ \Omega}$$
$$= 0.33$$

The answer is (A).

39. The effective isotropically radiated power (EIRP) is

$$\text{EIRP} = 10 \log G_T P_T$$

The subscript T indicates the transmitting antenna. The gain for the transmitting antenna is given in dBW, and must be converted to a numerical value as shown.

$$G_{T,\text{dBW}} = 10 \log G_T$$
$$17 = 10 \log G_T$$
$$1.7 = \log G_T$$
$$G_T = \text{antilog}\ 1.7 \approx 50$$

Substitute this value for the gain into the equation for the EIRP.

$$\text{EIRP} = 10 \log(G_T P_T) = 10 \log \frac{(50)(2\ \text{W})}{1\ \text{W}}$$
$$= 20\ \text{dBW}$$

The answer is (D).

40. The modulated FM signal, in terms of its constituent parts, is

$$s_{\text{mod}}(t) = A \cos\left[\omega_c t + \left(\frac{\Delta \omega}{\omega_{\text{mod}}}\right) \sin \omega_{\text{mod}} t\right]$$

Start with the index of modulation, m_{FM}, which is the term $\Delta\omega/\omega_{\text{mod}}$, and substitute the given information to solve for Δf.

$$m_{\text{FM}} = \frac{\Delta\omega}{\omega_{\text{mod}}}$$
$$8 = \frac{2\pi \Delta f}{10^6\ \frac{\text{rad}}{\text{s}}}$$
$$\Delta f = \frac{8 \times 10^6\ \frac{\text{rad}}{\text{s}}}{2\pi}$$
$$= 1.27 \times 10^6\ \text{Hz}\ \ (1.3 \times 10^6\ \text{Hz})$$

In terms of the coherent SI system, the unit "radian" has a dimension of 1. The term "rad" is used to aid understanding only.

The answer is (C).

Solutions
Afternoon Session Exam

41. The figure in option A is the piecewise linear model for a diode. The figure in option B is an h-parameter, small-signal, simplified equivalent circuit for a common-emitter bipolar junction transistor (BJT). The figure in option C is an h-parameter, small-signal, simplified equivalent circuit for a common-collector BJT. The figure in option D is a simplified model of a field-effect transistor. (The simplifications in each model consist of ignoring parameters that are insignificant; for example, extremely small reverse currents or near-infinite parallel resistance.)

The answer is (D).

42. The information provided is for the signal at V_2. To find V_1, apply Kirchhoff's current law at the negative terminal of the second-stage operational amplifier.

$$\frac{V^-(t) - V_2(t)}{240 \text{ k}\Omega} + \frac{V^-(t) - 0 \text{ V}}{60 \text{ k}\Omega} = 0$$
$$\frac{V^-(t) - V_2(t)}{240 \text{ k}\Omega} = -\frac{V^-(t)}{60 \text{ k}\Omega}$$
$$V^-(t) - V_2(t) = -4\,V^-(t)$$
$$V_2(t) = 5\,V^-(t)$$

Since the operational amplifier is ideal, or nearly so as is often the case, the negative terminal voltage equals the positive terminal voltage, in this case V_1. Making this substitution gives

$$V_2(t) = 5\,V_1(t)$$

Per the Fourier series, the second harmonic for V_2 is 1.5. From the above equation, the second harmonic for V_1 is one-fifth of that, or 0.3.

The answer is (A).

43. The voltage output of a linear variable differential transformer (LVDT) is shown here. Units shown are engineering units (eu). Various designs can have different full scale values, but in all designs 100 eu is the full scale value.

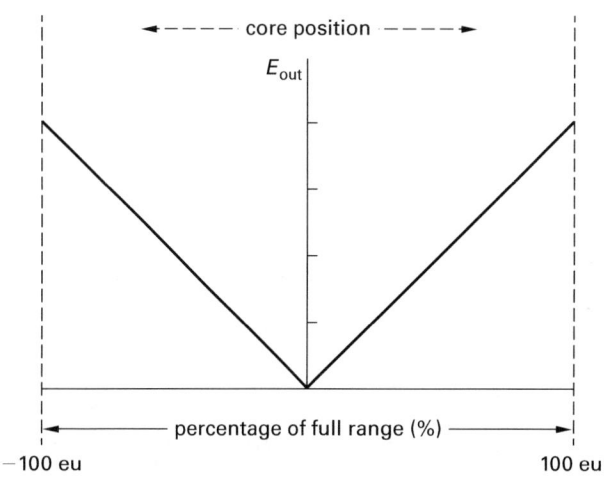

In this case, the full scale value is 100 eu. At a position, P, of 75 eu, the expected output voltage is

$$V_{\text{out}} = \left(\frac{P}{100 \text{ eu}}\right) V_{\text{full scale}}$$
$$= \left(\frac{75 \text{ eu}}{100 \text{ eu}}\right)(12 \text{ V})$$
$$= 9.0 \text{ V}$$

The error can be as large as 0.5% of the full scale value of 12 V, or 0.6 V. If the LVDT operates as expected, the voltage range should be 9.0 V $\pm$ 0.6 V, or 8.4 V to 9.6 V.

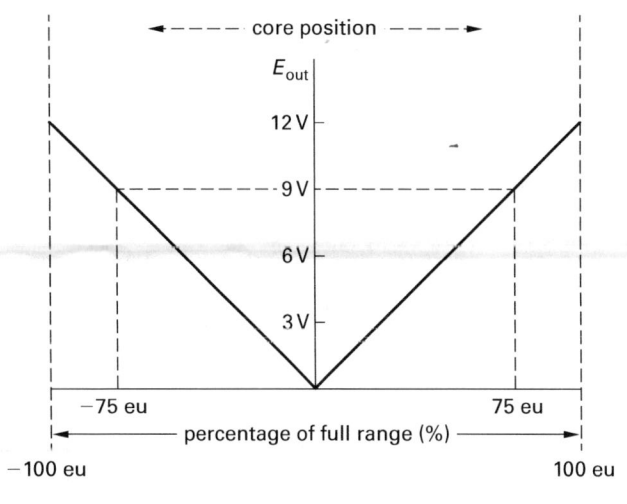

The answer is (B).

44. For both types of converter, the current source is the inductor and the voltage source is the capacitor within the circuitry, so statements I and II are true of both types.

For a boost converter, the transfer is called flyback energy transfer because the energy is stored in the magnetic field for use during the off cycle. The output voltage is higher than the input voltage and of the same polarity. Therefore, statements IV and VI are both true of a boost converter, and option D is correct.

For a buck converter, the transfer is called forward energy transfer because the energy is stored in the magnetic field and transferred during the on cycle. The output voltage is lower than the input voltage and of the same polarity. Therefore, statements III and V are both true of a buck converter, and options A and B are incorrect.

The answer is (D).

45. The equation for the current through the inductor is

$$v_L(t) = L\left(\frac{di(t)}{dt}\right)$$

Substituting the given current yields

$$v_L(t) = L\left(\frac{d(1.5 \text{ A})(1 - e^{-1000t})}{dt}\right)$$

$$= L\left(1.5\ \frac{\text{A}}{\text{s}}\right)(1000e^{-1000t})$$

$$= (5.0 \times 10^{-3}\text{ H})\left(1.5\ \frac{\text{A}}{\text{s}}\right)(1000e^{-1000t})$$

$$= 7.5e^{-1000t}\text{ V}$$

At 100 μs into a transient, the voltage is

$$v_L(t) = 7.5e^{-1000t}\text{ V}$$
$$= 7.5e^{-(1000)(100 \times 10^{-6}\text{ s})}\text{ V}$$
$$= 7.5e^{-0.1}\text{ V}$$
$$= 6.8\text{ V}$$

The answer is (B).

46. The relationship between the input and output voltages of a single-quadrant converter depends wholly upon the duty cycle. In the buck-boost configuration, however, the polarity of the output voltage is opposite that of the input. Use the relationship among the input and output voltages and duty cycle, and solve for the output.

$$\frac{V_{out}}{V_{in}} = \frac{-D}{1 - D}$$

$$V_{out} = -\left(\frac{D}{1 - D}\right)V_{in}$$

$$= -\left(\frac{0.8}{1 - 0.8}\right)(28\text{ V})$$

$$= -112\text{ V}\quad (-110\text{ V})$$

The answer is (A).

47. The electric field of the coaxial cable at $r = a$ is

$$\mathbf{E}_t = -\nabla\phi = \left(\frac{V_0}{\ln\dfrac{b}{a}}\right)\left(\frac{\mathbf{r}}{r}\right)$$

$$= \left(\frac{2000\text{ V}}{\ln\dfrac{4.95\text{ mm}}{0.91\text{ mm}}}\right)\left(\frac{\mathbf{r}}{(0.91\text{ mm})\left(1000\ \dfrac{\text{mm}}{\text{m}}\right)}\right)$$

$$= 1.298 \times 10^6\text{ V/m}\quad (1.3 \times 10^6\text{ V/m})$$

The answer is (C).

48. Find the impedance by dividing the voltage by the current.

$$Z = \frac{V}{I} = \frac{3 + j5\text{ V}}{4 - j10\text{ A}}$$

For complex numbers, the numerator and denominator are multiplied by the complex conjugate of the denominator.

$$Z = \left(\frac{3 + j5\text{ V}}{4 - j10\text{ A}}\right)\left(\frac{4 + j10}{4 + j10}\right)$$

$$= \frac{12 + j^2 50 + j20 + j30}{16 - j^2 100}\ \Omega$$

$$= \frac{12 - 50 + j50}{16 + 100}\ \Omega$$

$$= -0.33 + j0.43\ \Omega$$

The answer is (C).

49. Obtain the Thevenin equivalent of the gate-source circuit.

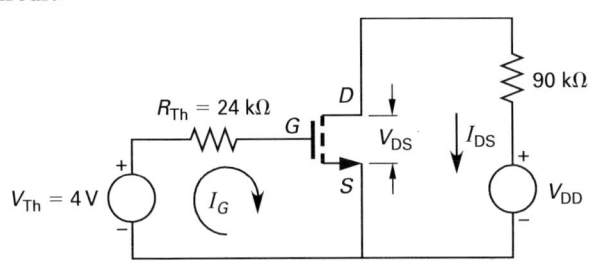

Write the KVL around the gate loop as indicated.

$$V_{Th} - I_G R_{Th} - V_{GS} = 0$$

The gate current in a MOSFET is zero amperes. Thus,

$$V_{GS} = V_{Th} = 4.0 \text{ V}$$

The answer is (D).

50. The critical incident angle, θ_c, for a wave incident upon a surface free space, which air approximates, is

$$\sin \theta_c = \frac{1}{n}$$

Rearrange and substitute the known value for the absolute index of refraction.

$$\begin{aligned}\theta_c &= \arcsin \frac{1}{n} \\ &= \arcsin \frac{1}{2} \\ &= 30° \quad (\pi/6 \text{ rad})\end{aligned}$$

The answer is (A).

51. The maximum frequency that can be carried is called the cutoff frequency. For a coaxial cable, the cutoff frequency is given by

$$\begin{aligned}f_c &= \frac{1.8412c}{2\pi a} \\ &= \frac{(1.8412)\left(3.00 \times 10^8 \ \frac{\text{m}}{\text{s}}\right)}{2\pi(14.1 \times 10^{-3} \text{ m})} \\ &= 6.23 \times 10^9 \text{ Hz} \quad (6.2 \text{ GHz})\end{aligned}$$

The answer is (D).

52. An open-loop transfer function for v_2 is one in which only the inputs (v_1, R_1, R_2) determine the output (v_2). From Ohm's law and Kirchhoff's laws,

$$v_{\text{load}} = iR_2$$
$$i = \frac{v_{\text{source}}}{R_1 + R_2}$$

Combining the two such that the load voltage is determined only by the inputs gives the open-loop transfer function.

$$v_{\text{load}} = \left(\frac{R_2}{R_1 + R_2}\right) v_{\text{source}}$$

Option D is the closed-loop function for v_{load} because the output is taken into account, that is, it is fed back to the input that determines the actual output.

The answer is (C).

53. Calculate the effective isotropic radiated power (EIRP) in decibel watts (dBW) as follows.

$$\begin{aligned}\text{EIRP} &= 10 \log \frac{G_T P_T}{1 \text{ W}} \\ &= 10 \log \frac{(5)(20{,}000 \text{ W})}{1 \text{ W}} \\ &= 50 \text{ dBW}\end{aligned}$$

Accounting for the cable losses of 1 dBW leaves an EIRP of 49 dBW. (The method used determines how cable losses are accounted. The power method incorporates the losses, and the linear method lists them separately. As losses are listed separately, they are already not accounted for in the values provided.)

The power at the receiving antenna is given by the following version of the Friss formula.

$$\begin{aligned}P_R &= \text{EIRP} + G_R - L_{\text{path}} \\ &= \text{EIRP} + G_R - (L_{\text{scatter}} + L_{\text{space}} + L_{\text{amp}}) \\ &= 49 \text{ dB} + 20 \text{ dB} - (3 \text{ dB} + 25 \text{ dB} + 1 \text{ dB}) \\ &= 40 \text{ dB}\end{aligned}$$

The answer is (C).

54. The gain of the antenna is the product of the efficiency, η, and the directivity, D.

$$G = \eta D$$

The directivity is calculated from the wavelength, λ, and the effective aperture, A_{eff}.

$$D = \left(\frac{4\pi}{\lambda^2}\right) A_{\text{eff}}$$

Because the total efficiency is given, the physical aperture measurement can be used (as opposed to the effective aperture, which already accounts for the aperture efficiency). The wavelength is calculated from the frequency.

$$D = \left(\frac{4\pi}{\left(\frac{c}{f}\right)^2}\right) A_{\text{phys}}$$

Substituting this into the equation for gain gives

$$G = \eta \left(\frac{4\pi}{\left(\frac{c}{f}\right)^2}\right) A_{\text{phys}}$$

$$= (0.7) \left(\frac{4\pi}{\left(\frac{3.00 \times 10^8 \, \frac{\text{m}}{\text{s}}}{2.4 \times 10^9 \, \text{Hz}}\right)^2}\right) (0.1 \text{ m}^2)$$

$$= 56.6 \quad (60)$$

The answer is (D).

55. One requirement for unconditional stability is that the magnitude of the source impedance must be less than that of the load impedance; in other words, the source-to-load impedance ratio must be less than one. The figure shows where this condition exists.

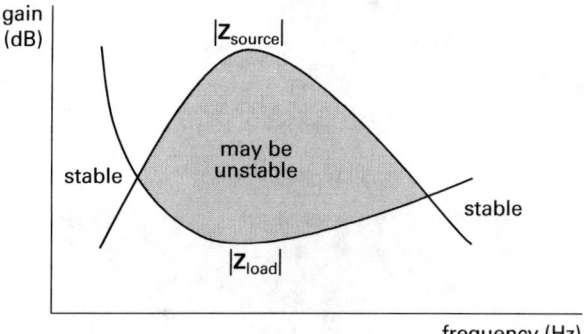

In the "may be unstable" region, further analysis must occur to determine stability.

The source-to-load impedance ratio is

$$R_{\text{stability}} = \frac{|\mathbf{Z}_{\text{source}}|}{|\mathbf{Z}_{\text{load}}|}$$

$$= \frac{|0.01 + j1.2 \, \Omega|}{|0.5 - j0.8 \, \Omega|}$$

$$= \frac{\sqrt{(0.01)^2 + (1.2)^2}}{\sqrt{(0.5)^2 + (-0.8)^2}}$$

$$= 1.272 \quad (1.3)$$

The source-to-load impedance ratio is greater than one, so it cannot be definitively stated that unconditional stability exists.

The answer is (D).

56. The collector-base cutoff current—essentially, the reverse saturation current—doubles with every 10°C rise. The thermal stability is affected by this current. The effect is self-reinforcing: as the temperature increases, the saturation current also increases. This is called thermal runaway. The emitter resistor helps stabilize the transistor against this trend. As the current rises, the voltage drop across the emitter resistor rises in a direction that opposes the forward biased base-emitter junction.

The answer is (A).

57. Control systems use the following form for the characteristic equation.

$$s^2 + 2\zeta\omega_n s + \omega_n^2$$

The damping ratio is represented by the symbol ζ. Solve for this value using the given information.

$$2\zeta\omega_n = 4$$

$$\zeta = \frac{4}{2\omega_n}$$

$$= \frac{4}{2\sqrt{144}}$$

$$= 0.17$$

The answer is (B).

58. Commercial AM broadcasts are generally of the double-sideband large carrier (DSB-LC) type—option D is true. Both sidebands are used—option C is true. These sidebands can be determined using Fourier analysis, which shows the signal as frequency shifted from the original carrier—option B is true.

Suppressing a carrier signal requires receivers to maintain synchronization and increases the cost, so the carrier signal remains in the commercial AM broadcasts.

The answer is (A).

59. All the polynomial terms are present, satisfying the Hurwitz criterion. The Routh table, constructed from the characteristic equation, is

s^3	1	3
s^2	6	C
s^1	$\dfrac{18-C}{6}$	0
s^0	C	

To be stable, all the entries in the first column must be positive. This requires that $0 < C < 18$. 11 is the only option that ensures system stability.

The answer is (C).

60. The thermal agitation noise, or Johnson noise, is

$$V_{\text{noise}} = \sqrt{4\kappa T R \Delta f}$$

The term κ is Boltzmann's constant, with a value of 1.3807×10^{-23} J/K. The term T is the absolute temperature, R is the resistance, and the bandwidth is represented by Δf. Substitute the given information, noting that 25°C is equal to 298K, to find the thermal agitation noise.

$$\begin{aligned} V_{\text{noise}} &= \sqrt{4\kappa T R \Delta f} \\ &= \sqrt{\begin{array}{c}(4)\left(1.3807\times 10^{-23}\,\dfrac{\text{J}}{\text{K}}\right)(298\text{K})\\ \times(20\times 10^3\,\Omega)(1000\text{ Hz})\end{array}} \\ &= 5.74\times 10^{-7}\text{ V}\quad(0.6\ \mu\text{V}) \end{aligned}$$

The answer is (D).

61. The curl of a given vector, **F**, can be found from determinant mathematics using the following general formula.

$$\begin{aligned} \text{curl }\mathbf{F} = \nabla\times\mathbf{F} &= \begin{vmatrix} \mathbf{i} & \mathbf{j} & \mathbf{k} \\ \dfrac{\partial}{\partial x} & \dfrac{\partial}{\partial y} & \dfrac{\partial}{\partial z} \\ F_x & F_y & F_z \end{vmatrix} \\ &= \left(\dfrac{\partial F_z}{\partial y}-\dfrac{\partial F_y}{\partial z}\right)\mathbf{i}+\left(\dfrac{\partial F_x}{\partial z}-\dfrac{\partial F_z}{\partial x}\right)\mathbf{j} \\ &\quad+\left(\dfrac{\partial F_y}{\partial x}-\dfrac{\partial F_x}{\partial y}\right)\mathbf{k} \end{aligned}$$

Substitute the given values for the magnetic field strength and apply the formula. Units are not shown, as is common practice within a matrix. (If the units of **H** are A/m, the units of **J** will be A/m^2.)

$$\begin{aligned} \mathbf{J}=\text{curl }\mathbf{H} &= \begin{vmatrix} \mathbf{i} & \mathbf{j} & \mathbf{k} \\ \dfrac{\partial}{\partial x} & \dfrac{\partial}{\partial y} & \dfrac{\partial}{\partial z} \\ (5x+3y) & (-5y-3z) & (5z-3x) \end{vmatrix} \\ &= \big(0-(-3)\big)\mathbf{i}+\big(0-(-3)\big)\mathbf{j}+(0-3)\mathbf{k} \\ &= 3\mathbf{i}+3\mathbf{j}-3\mathbf{k}\text{ A/m}^2 \end{aligned}$$

The answer is (A).

62. The output power can be found algebraically using the concept of decibels. First, express the input power in dBm.

$$\begin{aligned} P_{\text{in,dBm}} &= 10\log\dfrac{P_{\text{out}}}{P_{\text{ref}}} \\ &= 10\log\dfrac{30\text{ mW}}{1\text{ mW}} \\ &= 14.77\text{ dBm} \end{aligned}$$

This input power is reduced by the attenuator.

$$\begin{aligned} P_{c,\text{in,dBm}} &= P_{\text{in,dBm}}-P_{\text{loss,dB}} \\ &= 14.77\text{ dBm}-10\text{ dB} \\ &= 4.77\text{ dBm} \end{aligned}$$

The values with the ratios dBm and dB can be subtracted, since conversion of dB to dBm involves a factor of 10^{-3} in both the numerator and denominator. (See the calculation for P_{in}.)

The coupler splits the power. The power available at the meter is 3 dB less than the power at the coupler input, so

$$P_m = P_{c,\text{in,dBm}} - 3 \text{ dB}$$
$$= 4.77 \text{ dBm} - 3 \text{ dB}$$
$$= 1.77 \text{ dBm}$$

Convert to the unit requested.

$$P_{\text{in,dBm}} = 10 \log \frac{P_{\text{in,W}}}{P_{\text{ref}}}$$

$$P_{\text{in,W}} = \left(\text{antilog} \frac{P_{\text{in,dBm}}}{10}\right) P_{\text{ref}}$$

$$= \left(\text{antilog} \frac{1.77 \text{ dBm}}{10}\right)(1 \text{ mW})$$

$$= 1.50 \text{ mW}$$

The term dBm is unitless. It is shown as a reminder that the reference used for the decibel was milliwatts.

The answer is (A).

63. The thermal agitation noise is

$$V_{\text{noise}} = \sqrt{4\kappa TR(\text{BW})}$$

Convert the ambient temperature of 25°C to the absolute scale.

$$T_K = T_{°C} + 273.15°$$
$$= 25°C + 273.15°$$
$$= 298.15 \text{K}$$

Substitute into the noise formula, noting that κ is Boltzmann's constant.

$$V_{\text{noise}} = \sqrt{4\kappa TR(\text{BW})}$$

$$= \sqrt{\begin{array}{c}(4)\left(1.3805 \times 10^{-23} \dfrac{\text{J}}{\text{K}}\right)(298.15\text{K}) \\ \times (10 \times 10^3 \text{ }\Omega)(100 \times 10^3 \text{ Hz})\end{array}}$$

$$= 4.06 \times 10^{-6} \text{ V} \quad (4 \text{ }\mu\text{V})$$

The answer is (C).

64. The op amp responds to the instantaneous values of the voltage. The d'Arsonval meter movement responds to the root-mean-square voltage. The peak voltage associated with the maximum rms reading of 200 V is

$$|V_{\text{peak}}| = \sqrt{2} \, V_{\text{rms}}$$
$$= \sqrt{2} \, (200 \text{ V})$$
$$= 282.8 \text{ V}$$

The op amp maintains linear operation within the range ∓ 13 V. (The supply voltage is ± 16 V. Nominally, linear operation is maintained when the output voltage is at least 3 V from said voltages.) The op amp peak input voltage of 13 V (both positive and negative) should correspond to 282.8 V. The voltage divider attached to the positive terminal accomplishes this if the divider resistance is

$$(282.8 \text{ V})\left(\frac{R_{\text{div}}}{R_{\text{div}} + 1.5 \times 10^6 \text{ }\Omega}\right) = 13 \text{ V}$$

$$(282.8 \text{ V})R_{\text{div}} = (13 \text{ V})(R_{\text{div}} + 1.5 \times 10^6 \text{ }\Omega)$$

$$(282.8 \text{ V})R_{\text{div}} = (13 \text{ V})R_{\text{div}} + 19.50 \times 10^6 \text{ V·}\Omega$$

$$(282.8 \text{ V})R_{\text{div}} - (13 \text{ V})R_{\text{div}} = 19.50 \times 10^6 \text{ V·}\Omega$$

$$(269.8 \text{ V})R_{\text{div}} = 19.50 \times 10^6 \text{ V·}\Omega$$

$$R_{\text{div}} = \frac{19.50 \times 10^6 \text{ V·}\Omega}{269.8 \text{ V}}$$
$$= 7.23 \times 10^4 \text{ }\Omega \quad (0.07 \text{ M}\Omega)$$

Though the answer is rounded, high-precision resistors are used for scaling these types of op amp circuits.

The answer is (B).

65. The transistor is in saturation with the base-emitter and base-collector junctions forward biased. The first-order model is

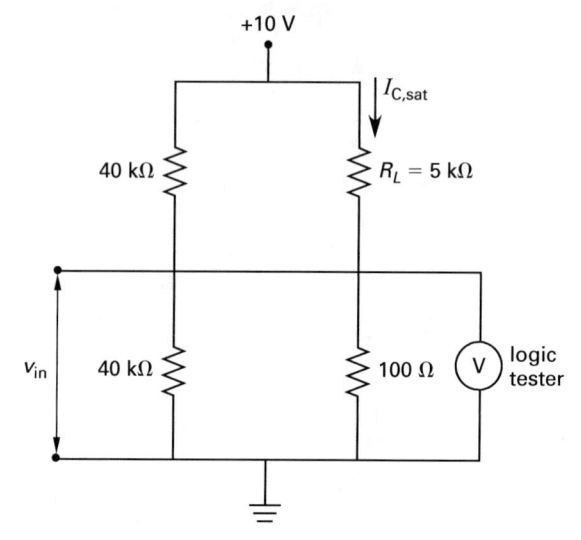

The current through the load resistor is the saturation current.

$$I_{C,\text{sat}} \approx \frac{V_{CC}}{R_L} = \frac{10 \text{ V}}{5 \times 10^3 \text{ }\Omega}$$
$$= 2 \times 10^{-3} \text{ A} \quad (2 \text{ mA})$$

The output voltage is low (near 0 V) when the transistor is saturated. The manufacturer's data sheet refers to this as "logic 1," so a low voltage is considered to be logic 1 (i.e., TRUE). The logic is "negative."

The answer is (C).

66. The 3 dB down point corresponds to the cutoff frequency as shown.

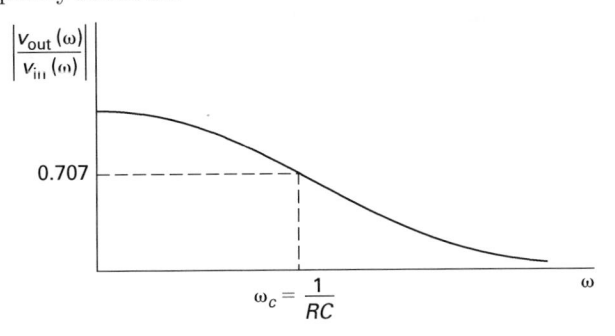

The cutoff frequency for the RC circuit given is

$$\omega_c = 2\pi f = \frac{1}{RC}$$

Substitute the given quantities and solve for the capacitance.

$$2\pi f = \frac{1}{RC}$$
$$C = \frac{1}{2\pi fR} = \frac{1}{2\pi (60 \text{ Hz})(50 \times 10^3 \ \Omega)}$$
$$= 5.31 \times 10^{-8} \text{ F} \quad (0.05 \ \mu\text{F})$$

Capacitor C_2 thus meets the stated requirement.

The answer is (B).

67. The circuit, a difference amplifier, is redrawn with parameters defined as

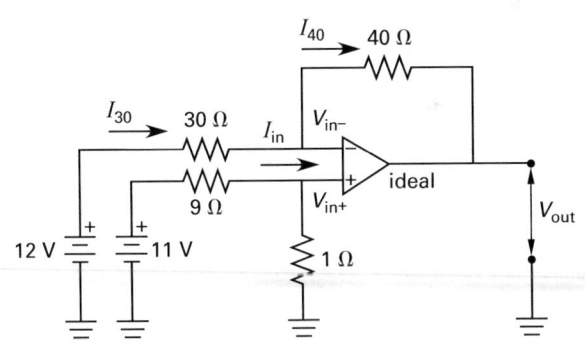

Using V_1 and the concept of a voltage divider, the voltage at the positive input is

$$V_{\text{in}^+} = V_1 \left(\frac{R_1}{R_1 + R_9} \right)$$
$$= (11 \text{ V}) \left(\frac{1 \ \Omega}{1 \ \Omega + 9 \ \Omega} \right)$$
$$= 1.1 \text{ V}$$

Given that ideal conditions may be applied, a virtual short circuit exists between the positive and negative inputs. That is, the voltages are identical (since the current through the feedback resistor, I_{40}, drives the value of V_{in^-} to the value of V_{in^+}). Thus,

$$V_{\text{in}^-} = V_{\text{in}^+}$$
$$= 1.1 \text{ V}$$

Apply Ohm's law to find the current through the 30 Ω resistor.

$$I_{30} = \frac{V_2 - V_{\text{in}^-}}{R_{30}}$$
$$= \frac{12 \text{ V} - 1.1 \text{ V}}{30 \ \Omega}$$
$$= 0.363 \text{ A}$$

Again, since the conditions are ideal, the input impedance is infinite and the input current is zero. Thus,

$$I_{\text{in}^-} = 0 \text{ A}$$
$$I_{30} = I_{40}$$
$$= 0.363 \text{ A}$$

Use Kirchhoff's voltage law in the feedback loop to find the output voltage.

$$V_{\text{out}} = V_{\text{in}^-} - I_{40}R_{40}$$
$$= 1.1 \text{ V} - (0.363 \text{ A})(40 \ \Omega)$$
$$= -13.420 \text{ V} \quad (-13 \text{ V})$$

The answer is (B).

68. The let-go current is approximately 10 mA.

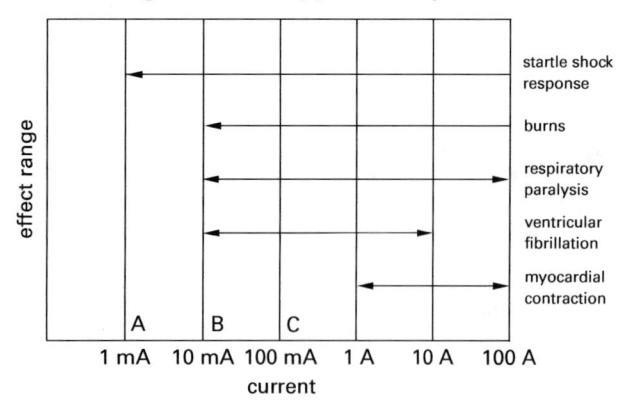

A = perception
B = let go
C = death

The voltage level required is

$$V = IR = (0.010 \text{ A})(500 \text{ }\Omega) = 5 \text{ V}$$

And, to generate enough current to potentially cause death (point C), 100 mA, would require only 50 V. The 50 V level is often used as the cutoff value for a low voltage circuit. Nevertheless, multiple definitions exist with differing requirements. For example, the *National Electrical Code* uses 24 V (per NEC Art. 551.2). Remote control circuits, often referred to as low voltage, are classified as those less than 50 V (per NEC Art. 720). Standards by the American National Standards Institute and the InterNational Electrical Testing Association (ANSI/NETA) define low voltage as less than 1000 V; ANSI/NETA standards are primarily concerned with distribution systems.

Older human body models used 300 Ω, and thus 30 V was the low voltage cutoff. In any case, very low AC voltage can result in harm to the human body. The body can be thought of as a capacitor, with water as the main dielectric. For this reason, AC voltage is more dangerous than DC voltage, which is unable to establish current flow at low voltages.

The answer is (A).

69. The transfer function is normally given as the response of the circuit divided by the forcing function, or $I(s)/V(s)$. Giving the function as $V(s)/I(s)$ puts it in a more standard form for determining poles and zeros.

The circuit in the *s*-domain is as shown.

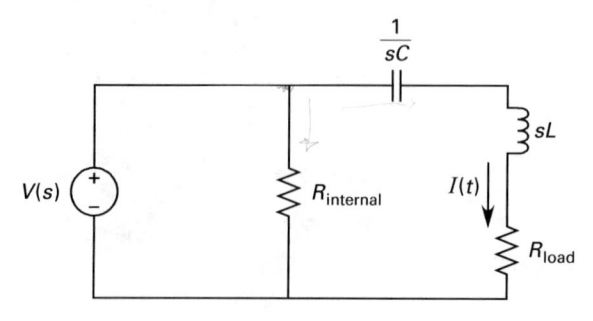

The transfer function in the *s*-domain can be found with the equation for a current divider.

$$I(s) = \left(\frac{R_{\text{internal}}}{\frac{1}{sC} + sL + R_{\text{load}}} \right) V(s)$$

$$\frac{I(s)}{V(s)} = \left(\frac{R_{\text{internal}}}{\frac{1}{sC} + sL + R_{\text{load}}} \right)$$

$$= \frac{4}{\frac{1}{0.1667s} + 2s + 8}$$

$$= \frac{4}{\frac{6}{s} + 2s + 8}$$

Multiplying top and bottom by $0.5s$ gives

$$\frac{I(s)}{V(s)} = \frac{2s}{3 + s^2 + 4s}$$

$$= \frac{2s}{(s+1)(s+3)}$$

The answer is (D).

70. The radio broadcast regulations are contained in the *Code of Federal Regulations*. (See Title 47, Chap. 73.14 (47CFR73.14), for an example of such regulations, which in this case sets the bandwidth allowed to AM stations.)

An AM receiver block diagram follows.

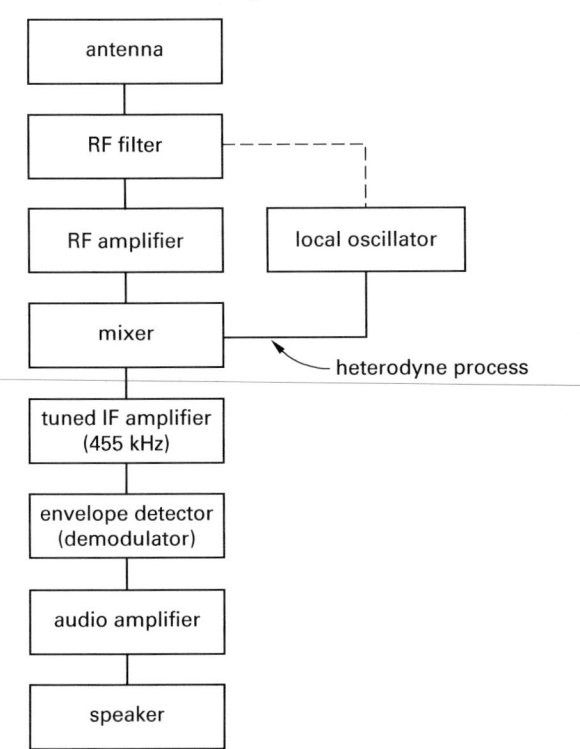

```
        antenna
           │
        RF filter ─ ─ ─ ─ ┐
           │              │
        RF amplifier   local oscillator
           │              │
         mixer ───────────┘
           │     ↖ heterodyne process
     tuned IF amplifier
        (455 kHz)
           │
    envelope detector
      (demodulator)
           │
      audio amplifier
           │
         speaker
```

The IF amplifier is tuned to 455 kHz. This is the intermediate frequency (IF) used in the heterodyne process. Tuning to an AM station at 680 kHz requires that the local oscillator be set for

$$f_{osc} = f_{station} + f_{IF}$$
$$= 680 \text{ kHz} + 455 \text{ kHz}$$
$$= 1135 \text{ kHz}$$

Only the higher of the two frequencies generated in the heterodyne process is used. That is, the heterodyne frequency 225 kHz is not used, since to do so would require the oscillator to tune to a wider band, proportionally.

The answer is (D).

71. The power density of the signal is

$$p_r = \left(\frac{D_T}{4\pi r^2}\right) P_T$$
$$= \left(\frac{1.4}{4\pi (5 \times 10^3 \text{ m})^2}\right)(5 \times 10^3 \text{ W})$$
$$= 22.28 \times 10^{-6} \text{ W/m}^2 \quad (22 \text{ }\mu\text{W/m}^2)$$

The answer is (A).

72. Carrier sense multiple access (CSMA) is a network protocol that verifies the absence of traffic before transmitting upon the given medium. There are multiple strategies within CSMA, including 1-persistent, non-persistent, O-persistent, and P-persistent. This is combination 1C.

Time-division multiple access (TDMA) is an older technique that allows users to share different time slots on the same data channel. This is combination 2A.

Fiber optic networks use wavelength-division multiplexing (WDM), sending different colors of light on the same channel. The multiplexing can occur as coarse WDM or dense WDM. This is combination 3B.

The answer is (D).

73. Spread-spectrum modulation takes a narrowband signal and spreads it over a wideband frequency range. Advantages include increased security and greater resistance to various kinds of interference.

The two techniques primarily employed are frequency-hopping spread spectrum (FHSS) and direct-sequence spread spectrum (DSSS). The DSSS technique uses a single frequency along with a secret code, called the chirping code, to vary the data. The FHSS technique uses multiple frequencies, hopping from one frequency to another at short, regular intervals according to a coded sequence called the spreading code.

FHSS is not impacted by interference, while DSSS can be. Option C is correct.

FHSS is more expensive to implement than DSSS, so option A is incorrect. Frequencies can be reused during FHSS data transmission, so option B is incorrect. The signal strength used in FHSS is higher than in the DSSS method, so option D is incorrect.

The answer is (C).

74. By convention, information rates are given in hertz (Hz), which in this context means samples per second.

The Nyquist rate is the rate that achieves an ideal sample of the signal. This rate is

$$f_s = 2f_i = (2)\left(3000 \text{ }\frac{\text{samples}}{\text{s}}\right)$$
$$= 6000 \text{ samples/s} \quad (6000 \text{ Hz})$$

In practice, the rate should be higher than this value in order to avoid aliasing, or foldover, which distorts the original signal.

The answer is (D).

75. The reliability of an item is the probability that it will not fail before a given time. As this component will fail only by random causes and does not experience deterioration during its life, its reliability is calculated from the mean time to failure, MTTF, as

$$R(t) = e^{-t/\text{MTTF}}$$

$$= \exp\left[\frac{-(3 \text{ yr})\left(365 \frac{\text{days}}{\text{yr}}\right)\left(24 \frac{\text{hr}}{\text{day}}\right)}{40{,}000 \text{ hr}}\right]$$

$$= 0.518 \quad (0.5)$$

The mean time to failure is also known as the mean time before failure, MTBF.

The answer is (B).

76. Per the *National Electric Code* (NEC) Table 250.122, the grounding conductor is required to be 12 AWG in size.

The answer is (C).

77. The testing provides the open-circuit, or Thevenin, voltage and the short-circuit, or Thevenin, current.

$$\mathbf{V}_{\text{Th}} = \mathbf{V}_{oc} = 28 \text{ V}\angle 0°$$
$$\mathbf{I}_{\text{Th}} = \mathbf{I}_{sc} = 7 \text{ A}\angle 80°$$

The Thevenin impedance is

$$\mathbf{Z}_{\text{Th}} = \frac{\mathbf{V}_{\text{Th}}}{\mathbf{I}_{\text{Th}}}$$
$$= \frac{28 \text{ V}\angle 0°}{7 \text{ A}\angle 80°}$$
$$= 4 \text{ }\Omega\angle -80°$$

In rectangular form,

$$\mathbf{Z}_{\text{Th}} = Z_{\text{Th}} \cos\phi + jZ_{\text{Th}} \sin\phi$$
$$= (4 \text{ }\Omega)\cos(-80°)$$
$$\quad + j(4 \text{ }\Omega)\sin(-80°)$$
$$= 0.6946 - j3.939 \text{ }\Omega$$
$$(0.69 - j3.9 \text{ }\Omega)$$

The answer is (C).

78. Using the right-hand rule, placing the thumb in the direction of the current flow, the fingers curl inward toward the direction of the magnetic field. Option D illustrates the result.

The answer is (D).

79. The Karnaugh map one-squares can be grouped as follows, allowing the "don't care" condition in m_3 to take on the value of 1.

C \ AB	00	01	11	10
0	1	0	d	1
1	1	d	1	1

$$F(A,B,C) = \overline{B} + C$$

The function $F(A, B, C)$ is realized with the PLA given by option D.

The answer is (D).

80. The sum-of-products (SOP) form of the output function uses the minterms of the function. Minterms are those terms in which the value of the function is 1, or true. (Maxterms take on the value of 0.) The minterms, m, and maxterms, M, for the given function are shown.

X	Y	Z	$F(X, Y, Z)$	minterms, m maxterms, M
0	0	0	0	M_0
0	0	1	1	m_1
0	1	0	0	M_2
0	1	1	1	m_3
1	0	0	0	M_4
1	0	1	1	m_5
1	1	0	0	M_6
1	1	1	1	m_7

Minterms exist at input values of one, three, five, and seven. The function, in canonical SOP form, is

$$F(X, Y, Z) = m_1 + m_3 + m_5 + m_7$$
$$= \overline{X}\,\overline{Y}Z + \overline{X}YZ + X\overline{Y}Z + XYZ$$

The answer is (A).